ORIGAMI, ELEUSIS, AND THE SOMA CUBE

For 25 of his 90 years, Martin Gardner wrote "Mathematical Games and Recreations," a monthly column for *Scientific American* magazine. These columns have inspired hundreds of thousands of readers to delve more deeply into the large world of mathematics. He has also made significant contributions to magic, philosophy, debunking pseudoscience, and children's literature. He has produced more than 60 books, including many best sellers, most of which are still in print. His *Annotated Alice* has sold more than a million copies. He continues to write a regular column for the *Skeptical Inquirer* magazine. (The photograph is of the author in 1959.)

THE NEW MARTIN GARDNER MATHEMATICAL LIBRARY

From 1957 through 1986 Martin Gardner wrote the "Mathematical Games" columns for *Scientific American* that are the basis for these books. *Scientific American* editor Dennis Flanagan noted that this column contributed substantially to the success of the magazine. The exchanges between Martin Gardner and his readers gave life to these columns and books. These exchanges have continued and the impact of the columns and books has grown. These new editions give Martin Gardner the chance to bring readers up to date on newer twists on old puzzles and games, on new explanations and proofs, and on links to recent developments and discoveries. Illustrations have been added and existing ones improved, and the bibliographies have been greatly expanded throughout.

1. Hexaflexagons, Probability Paradoxes, and the Tower of Hanoi: Martin Gardner's First Book of Mathematical Puzzles and Games
2. Origami, Eleusis, and the Soma Cube: Martin Gardner's Mathematical Diversions
3. Sphere Packing, Lewis Carroll, and Reversi: Martin Gardner's New Mathematical Diversions
4. Knots and Borromean Rings, Rep-Tiles, and Eight Queens: Martin Gardner's Unexpected Hanging
5. Klein Bottles, Op-Art, and Sliding-Block Puzzles: More of Martin Gardner's Mathematical Games
6. Sprouts, Hypercubes, and Superellipses: Martin Gardner's Mathematical Carnival

Origami, Eleusis, and the Soma Cube

MARTIN GARDNER'S MATHEMATICAL DIVERSIONS

Martin Gardner

The Mathematical Association of America

CAMBRIDGE
UNIVERSITY PRESS

32 Avenue of the Americas, New York NY 10013-2473, USA

Cambridge University Press is part of the University of Cambridge.

It furthers the University's mission by disseminating knowledge in the pursuit of education, learning and research at the highest international levels of excellence.

www.cambridge.org
Information on this title: www.cambridge.org/9780521735247

© Mathematical Association of America 2008

First published 2008

A catalogue record for this publication is available from the British Library

Library of Congress Cataloguing in Publication data

Gardner, Martin, 1914–
Origami, Eleusis, and the Soma cube : Martin Gardner's mathematical
diversions / Martin Gardner.
p. cm. – (The new Martin Gardner mathematical library)
Includes bibliographical references and index.
ISBN 978-0-521-75610-5 (hardback)
1. Mathematical recreations. I. Title. II. Series.
QA95.G2975 2008
793.74–dc22 2008012534

ISBN 978-0-521-75610-5 Hardback
ISBN 978-0-521-73524-7 Paperback

Contents

Acknowledgments

Martin Gardner thanks *Scientific American* for allowing reuse of material from his columns in that magazine, material copyright © 1958 (Chapters 1–7, 9, 17), 1959 (Chapters 8, 10–16, 18, 19), and 1960 (Chapter 20) by Scientific American, Inc. He also thanks the artists who contributed to the success of these columns and books for allowing reuse of their work: James D. Egelson (via heirs Jan and Nicholas Egelson), Harold Jacobs, Amy Kasai, Alex Semenoick, and Bunji Tagawa (via Donald Garber for the Tagawa Estate). Artists' names are cited where these were known. All rights other than use in connection with these materials lie with the original artists.

Photograph of Bernardino Luini, "Boy with a Toy," Elton Hall Collection is reproduced from "Gibeciere" vol. 1, no. 1 and used by permission. Photograph in Figure 51, courtesy of National Gallery of Art, copyright the Salvador Dali estate, is used by permission. Photograph of Albrecht Durer's Melancolia I is courtesy of Owen Gingerich. Photograph in figure 36 is courtesy of Ed Vogel.

Introduction

Since the appearance of the first *Scientific American Book of Mathematical Puzzles & Diversions*, in 1959, popular interest in recreational mathematics has continued to increase. Many new puzzle books have been printed, old puzzle books have been reprinted, kits of recreational math materials are on the market, a new topological game (see Chapter 7) has caught the fancy of the country's youngsters, and an excellent little magazine called *Recreational Mathematics* has been started by Joseph Madachy, a research chemist in Idaho Falls. Chessmen – those intellectual status symbols – are jumping all over the place, from TV commercials and magazine advertisements to Al Horowitz's lively chess corner in *The Saturday Review* and the knight on Paladin's holster and have-gun-will-travel card.

This pleasant trend is not confined to the United States. A classic four-volume French work, *Récréations Mathématiques,* by Edouard Lucas, has been reissued in France in paperback. Thomas H. O'Beirne, a Glasgow mathematician, is writing a splendid puzzle column in a British science journal. A handsome 575-page collection of puzzles, assembled by mathematics teacher Boris Kordemski, is selling in Russian and Ukrainian editions. It is all, of course, part of a worldwide boom in math – in turn a reflection of the increasing demand for skilled mathematicians to meet the incredible needs of the new triple age of the atom, spaceship, and computer.

Computers are not replacing mathematicians; they are breeding them. It may take a computer less than 20 seconds to solve a thorny problem, but it may have taken a group of mathematicians many months to program the problem. In addition, scientific research is becoming more and more dependent on the mathematician for important breakthroughs in theory. The relativity revolution, remember, was the work of a man who had no experience in the laboratory. At the moment, atomic scientists are thoroughly befuddled by the preposterous properties of some 30 different fundamental particles, "a vast jumble of odd dimensionless numbers," as J. Robert Oppenheimer has described them, "none of them understandable or derivable, all with an insulting lack of obvious meaning." One of these days a great creative mathematician, sitting alone and scribbling on a piece of paper, or shaving, or taking his family on a picnic, will experience a flash of insight. The particles will spin into their appointed places, rank on rank, in a beautiful pattern of unalterable law. At least, that is what the particle physicists *hope* will happen. Of course the great puzzle solver will draw on laboratory data, but the chances are that he will be, like Einstein, primarily a mathematician.

Not only in the physical sciences is mathematics battering down locked doors. The biological sciences, psychology, and the social sciences are beginning to reel under the invasion of mathematicians armed with strange new statistical techniques for designing experiments, analyzing data, and predicting probable results. It may still be true that if the president of the United States asks three economic advisers to study an important question, they will report back with four different opinions, but it is no longer absurd to imagine a distant day when economic disagreements can be settled by mathematics in a way that is not subject to the usual dismal disputes. In the cold light of modern economic theory, the conflict between socialism and capitalism is rapidly becoming, as Arthur Koestler has put it, as naïve and sterile as the wars in Lilliput over the two ways to break an egg. (I speak only of the economic debate; the conflict between democracy and totalitarianism has nothing to do with mathematics.)

But those are weighty matters, and this is only a book of amusements. If it has any serious purpose at all, it is to stimulate popular interest in mathematics. Such stimulation is surely desirable, if for no other reason than to help the layman understand what the scientists are up to. And they are up to plenty.

I would like to express again my gratitude to the publisher, editors, and staff of *Scientific American*, the magazine in which these chapters first appeared; to my wife for assistance in many ways; and to the hundreds of friendly readers who continue to correct my errors and suggest new material. I would like also to thank, for her expert help in preparing the manuscript, Nina Bourne of Simon and Schuster.

MARTIN GARDNER

CHAPTER ONE

The Five Platonic Solids

A REGULAR POLYGON is a plane figure bounded by straight lines, with equal sides and equal interior angles. There is of course an infinite number of such figures. In three dimensions the analog of the regular polygon is the regular polyhedron: a solid bounded by regular polygons, with congruent faces and congruent interior angles at its corners. One might suppose that these forms are also infinite, but in fact they are, as Lewis Carroll once expressed it, "provokingly few in number." There are only five regular convex solids: the regular tetrahedron, hexahedron (cube), octahedron, dodecahedron, and icosahedron (see Figure 1).

The first systematic study of the five regular solids appears to have been made by the ancient Pythagoreans. They believed that the tetrahedron, cube, octahedron, and icosahedron respectively underlay the structure of the traditional four elements: fire, earth, air, and water. The dodecahedron was obscurely identified with the entire universe. Because these notions were elaborated in Plato's *Timaeus*, the regular polyhedrons came to be known as the Platonic solids. The beauty and fascinating mathematical properties of these five forms haunted scholars from the time of Plato through the Renaissance. The analysis of the Platonic solids provides the climactic final book of Euclid's *Elements*. Johannes Kepler believed throughout his life that the orbits of the six planets known in his day could be obtained by nesting the five solids in a certain order within the orbit of Saturn. Today the mathematician no longer views the Platonic solids with mystical reverence, but their rotations are studied in connection with group theory, and they continue to play a colorful role in recreational mathematics. Here we shall quickly examine a few diversions in which they are involved.

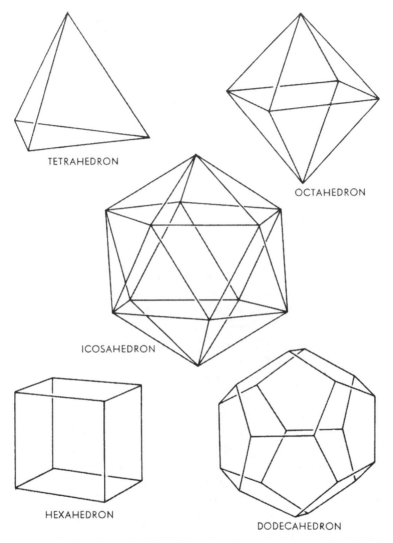

TETRAHEDRON

OCTAHEDRON

ICOSAHEDRON

HEXAHEDRON

DODECAHEDRON

Figure 1. The five Platonic solids. The cube and octahedron are "duals" in the sense that if the centers of all pairs of adjacent faces on one are connected by straight lines, the lines form the edges of the other. The dodecahedron and icosahedron are dually related in the same way. The tetrahedron is its own dual. (Artist: Bunji Tagawa)

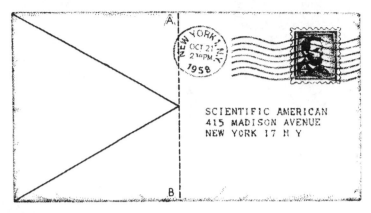

Figure 2. How a sealed envelope can be cut for folding into a tetrahedron. (Artist: Bunji Tagawa)

There are four different ways in which a sealed envelope can be cut and folded into a tetrahedron. The following is perhaps the simplest. Draw an equilateral triangle on both sides of one end of an envelope (see Figure 2). Then cut through both layers of the envelope as indicated by the broken line and discard the right-hand piece. By creasing the paper along the sides of the front and back triangles, points A and B are brought together to form the tetrahedron.

Figure 3 shows the pattern for a tantalizing little puzzle currently marketed in plastic. You can make the puzzle yourself by cutting two such patterns out of heavy paper. (All the line segments except the longer one have the same length.) Fold each pattern along the lines and tape the edges to make the solid shown. Now try to fit the two solids together to make a tetrahedron. A mathematician I know likes to annoy his friends with a practical joke based on this puzzle. He bought two sets of the plastic pieces so that he could keep a third piece concealed in his hand. He displays a tetrahedron on the table, then knocks it over with his hand and at the same time releases the concealed piece. Naturally his friends do not succeed in forming the tetrahedron out of the *three* pieces.

Concerning the cube, I shall mention only an electrical puzzle and the surprising fact that a cube can be passed through a hole in a smaller cube. If you will hold a cube so that one corner points

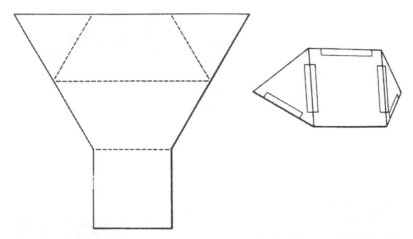

Figure 3. A pattern (left) that can be folded into a solid (right), two of which make a tetrahedron. (Artist: Bunji Tagawa)

directly toward you, the edges outlining a hexagon, you will see at once that there is ample space for a square hole that can be slightly larger than the face of the cube itself. The electrical puzzle involves the network depicted in Figure 4. If each edge of the cube has a resistance of one ohm, what is the resistance of the entire structure when current flows from A to B? Electrical engineers have been known to produce pages of computations on this problem, though it yields easily to the proper insight.

All five Platonic solids have been used as dice. Next to the cube the octahedron seems to have been the most popular. The pattern shown in Figure 5, its faces numbered as indicated, will fold into a

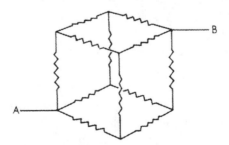

Figure 4. An electrical-network puzzle. (Artist: Bunji Tagawa)

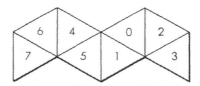

Figure 5. A strip to make an octahedral die. (Artist: Bunji Tagawa)

neat octahedron whose open edges can be closed with transparent tape. The opposite sides of this die, as in the familiar cubical dice, total seven. Moreover, a pleasant little mind-reading stunt is made possible by this arrangement of digits. Ask someone to think of a number from 0 to 7 inclusive. Hold up the octahedron so that he sees only the faces 1, 3, 5, and 7, and ask him if he sees his chosen number. If he says "Yes," this answer has a key value of 1. Turn the solid so that he sees faces 2, 3, 6, and 7, and ask the question again. This time "Yes" has the value of 2. The final question is asked with the solid turned so that he sees 4, 5, 6, and 7. Here a "Yes" answer has the value of 4. If you now total the values of his three answers you obtain the chosen number, a fact that should be easily explained by anyone familiar with the binary system. To facilitate finding the three positions in which you must hold the solid, simply mark in some way the three corners that must be pointed toward you as you face the spectator.

There are other interesting ways of numbering the faces of an octahedral die. It is possible, for example, to arrange the digits 1 through 8 in such a manner that the total of the four faces around each corner is a constant. The constant must be 18, but there are three distinct ways (not counting rotations and reflections) in which the faces can be numbered in this fashion.

An elegant way to construct a dodecahedron is explained in Hugo Steinhaus's book *Mathematical Snapshots*. Cut from heavy cardboard two patterns like the one pictured at left in Figure 6. The pentagons should be about an inch on a side. Score the outline of each center pentagon with the point of a knife so that the pentagon flaps fold easily in one direction. Place the patterns together as shown at right in the illustration so that the flaps of each pattern fold toward the others. Weave a rubber band alternately over

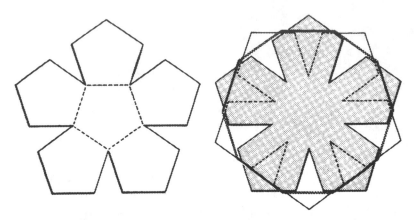

Figure 6. Two identical patterns are fastened together with a rubber band to make a pop-up dodecahedron. (Artist: Bunji Tagawa)

and under the projecting ends, keeping the patterns pressed flat. When you release the pressure, the dodecahedron will spring magically into shape.

If the faces of this model are colored, a single color to each face, what is the minimum number of colors needed to make sure that no edge has the same color on both sides? The answer is four, and it is not difficult to discover the four different ways that the colors can be arranged (two are mirror images of the other two). The tetrahedron also requires four colors, there being two arrangements, one a reflection of the other. The cube needs three colors and the octahedron two, each having only one possible arrangement. The icosahedron calls for three colors; here there are no less than 144 different patterns, only six of which are identical with their mirror images.

If a fly were to walk along the 12 edges of an icosahedron, traversing each edge at least once, what is the shortest distance it could travel? The fly need not return to its starting point, and it would be necessary for it to go over some edges twice. (Only the octahedron's edges can be traversed without retracing.) A plane projection of the icosahedron (Figure 7) may be used in working on this problem, but one must remember that each edge is one unit in length. (I have been unable to resist concealing a laconic Christmas greeting in the

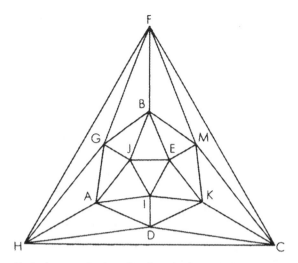

Figure 7. A plane projection of an icosahedron. (Artist: Bunji Tagawa)

way the corners of this diagram are labeled. It is not necessary to solve the problem in order to find it.)

In view of the fact that cranks persist in trying to trisect the angle and square the circle long after these feats have been proved impossible, why has there been no comparable effort to find more than five regular polyhedrons? One reason is that it is quite easy to "see" that no more are possible. The following simple proof goes back to Euclid.

A corner of a polyhedron must have at least three faces. Consider the simplest face: an equilateral triangle. We can form a corner by putting together three, four, or five such triangles. Beyond five, the angles total 360 degrees or more and therefore cannot form a corner. We thus have three possible ways to construct a regular convex solid with triangular faces. Three and only three squares will similarly form a corner, indicating the possibility of a regular solid with square faces. The same reasoning yields one possibility with three pentagons at each corner. We cannot go beyond the pentagon, because when we put three hexagons together at a corner, they equal 360 degrees.

This argument does not prove that five regular solids can be constructed, but it does show clearly that no more than five are possible. More sophisticated arguments establish that there are six regular polytopes, as they are called, in four-dimensional space. Curiously, in every space of more than four dimensions there are only three regular polytopes: analogs of the tetrahedron, cube, and octahedron.

A moral may be lurking here. There is a very real sense in which mathematics limits the kinds of structures that can exist in nature. It is not possible, for example, that beings in another galaxy gamble with dice that are regular convex polyhedra of a shape unknown to us. Some theologians have been so bold as to contend that not even God himself could construct a sixth Platonic solid in three-dimensional space. In similar fashion, geometry imposes unbreakable limits on the varieties of crystal growth. Some day physicists may even discover mathematical limitations to the number of fundamental particles and basic laws. No one of course has any notion of how mathematics may, if indeed it does, restrict the nature of structures that can be called "alive." It is conceivable, for example, that the properties of carbon compounds are absolutely essential for life. In any case, as humanity braces itself for the shock of finding life on other planets, the Platonic solids serve as ancient reminders that there may be fewer things on Mars and Venus than are dreamt of in our philosophy.

ANSWERS

The total resistance of the cubical network is 5/6 ohm. If the three corners closest to A are short-circuited together, and the same is done with the three corners closest to B, no current will flow in the two triangles of short circuits because each connects equipotential points. It is now easy to see that there are three one-ohm resistors in parallel between A and the nearest triangle (resistance 1/3 ohm), six in parallel between the triangles (1/6 ohm), and three in parallel between the second triangle and B (1/3 ohm), making a total resistance of 5/6 ohm.

C. W. Trigg, discussing the cubical-network problem in the November–December 1960 issue of *Mathematics Magazine*, points out that a solution for it may be found in *Magnetism and Electricity*, by E. E. Brooks and A. W. Poyser, 1920. The problem and the method of solving it can be easily extended to networks in the form of the other four Platonic solids.

The three ways to number the faces of an octahedron so that the total around each corner is 18 are 6, 7, 2, 3 clockwise (or counter-clockwise) around one corner, and 1, 4, 5, 8 around the opposite corner (6 adjacent to 1, 7 to 4, and so on); 1, 7, 2, 8 and 4, 6, 3, 5; and 4, 7, 2, 5 and 6, 1, 8, 3. See W. W. Rouse Ball's *Mathematical Recreations and Essays,* Chapter 7, for a simple proof that the octahedron is the only one of the five solids whose faces can be numbered so that there is a constant sum at each corner.

The shortest distance the fly can walk to cover all edges of an icosahedron is 35 units. By erasing five edges of the solid (for example, edges FM, BE, JA, ID, and HC) we are left with a network that has only two points, G and K, where an odd number of edges come together. The fly can therefore traverse this network by starting at G and going to K without retracing an edge – a distance of 25 units. This is the longest distance it can go without retracing. Each erased edge can now be added to this path, whenever the fly reaches it, simply by traversing it back and forth. The five erased edges, each gone over twice, add 10 units to the path, making a total of 35.

POSTSCRIPT

Margaret Wertheim, writing on "A Puzzle Finally Makes the 'Cosmic Figures' Fit," in *The New York Times* (May 10, 2005), describes a remarkable puzzle created by Dr. Wayne Daniel, a retired physicist living in Genoa, Nevada. Called All Five, it consists of 41 wooden pieces that form the five Platonic solids, all nested together like Russian matryoshka dolls. Outside is the icosahedron, followed by the dodecahedron, cube, tetrahedron, and in the center a tiny octahedron. There are no empty spaces between the pieces! Dr. Daniel has constructed other puzzles based on the five regular solids, but this one is his crowning achievement. He has made a DVD showing how the pieces come apart and go back together. It can be seen on his Web site.

In the books to follow in this series, there are many references to problems and curiosities involving the five solids. Note in particular a chapter in Book 10 on Jean Pedersen's way of plaiting polyhedra with paper strips, and references there cited. A chapter devoted entirely to tetrahedra is in Book 5.

One can imagine how amazed and delighted Plato and Kepler would have been if someone had given them an All Five.

BIBLIOGRAPHY

"Folding an Envelope into Tetrahedra." C. W. Trigg in *The American Mathematical Monthly* 56:6 (June–July 1949): 410–412.

Mathematical Models. H. Martyn Cundy and A. P. Rollett. Clarendon Press, 1952.

"Geometry of Paper Folding II: Tetrahedral Models." C. W. Trigg in *School Science and Mathematics* (December 1954): 683–689.

"The Perfect Solids." Arthur Koestler in *The Watershed*, a biography of Johannes Kepler. Chapter 2. Doubleday Anchor Books, 1960. An excellent discussion of Kepler's attempt to explain the planetary orbits by means of the Platonic solids.

CHAPTER TWO

Tetraflexagons

HEXAFLEXAGONS are diverting six-sided paper structures that can be "flexed" to bring different surfaces into view. They are constructed by folding a strip of paper as explained in Book 1. Close cousins to the hexaflexagons are a wide variety of four-sided structures that may be grouped loosely under the term "tetraflexagon."

Hexaflexagons were invented in 1939 by Arthur H. Stone, then a graduate student at Princeton University and now a lecturer in mathematics at the University of Manchester in England. Their properties have been thoroughly investigated; indeed, a complete mathematical theory of hexaflexigation has been developed. Much less is known about tetraflexagons. Stone and his friends (notably John W. Tukey, now a well-known topologist) spent considerable time folding and analyzing these four-sided forms, but they did not succeed in developing a comprehensive theory that would cover all their discordant variations. Several species of tetraflexagon are nonetheless intensely interesting from the recreational standpoint.

Consider first the simplest tetraflexagon, a three-faced structure that can be called the tri-tetraflexagon. It is easily folded from the strip of paper shown in Figure 8 (8a is the front of the strip; 8b, the back). Number the small squares on each side of the strip as indicated, fold both ends inward (8c) and join two edges with a piece of transparent tape (8d). Face 2 is now in front; face 1 is in back. To flex the structure, fold it back along the vertical center line of face 2. Face 1 will fold into the flexagon's interior as face 3 flexes into view.

Stone and his friends were not the first to discover this interesting structure; it has been used for centuries as a double-action hinge. I have on my desk, for instance, two small picture frames containing

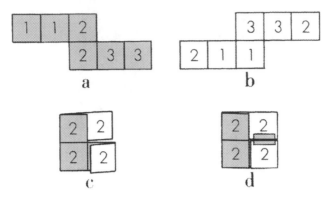

Figure 8. How to make a tri-tetraflexagon. (Artist: Bunji Tagawa)

photographs. The frames are joined by two tri-tetraflexagon hinges which permit the frames to flex forward or backward with equal ease.

The same structure is involved in several children's toys, the most familiar of which is a chain of flat wooden or plastic blocks hinged together with crossed tapes. If the toy is manipulated properly, one block seems to tumble down the chain from top to bottom. Actually this is an optical illusion created by the flexing of the tri-tetraflexagon hinges in serial order. The toy was popular in the United States during the 1890s, when it was called Jacob's Ladder. (A picture and description of the toy appear in Albert A. Hopkins's *Magic: Stage Illusions and Scientific Diversions,* 1897.) Two current models sell under the trade names Klik-Klak Blox and Flip Flop Blocks.

There are at least six types of four-faced tetraflexagons, known as tetra-tetraflexagons. A good way to make one is to start with a rectangular piece of thin cardboard ruled into 12 squares. Number the squares on both sides as depicted in Figure 9 (9a and 9b). Cut the rectangle along the broken lines. Start as shown in 9a, then fold the two center squares back and to the left. Fold back the column on the extreme right. The cardboard should now appear as shown in 9c. Again fold back the column on the right. The single square projecting on the left is now folded forward and to the right. This brings all six of the "1" squares to the front. Fasten together the edges of the two middle squares with a piece of transparent tape as shown in 9d.

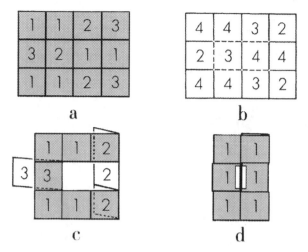

Figure 9. How to make a tetra-tetraflexagon. (Artist: Bunji Tagawa)

You will find it a simple matter to flex faces 1, 2, and 3 into view, but finding face 4 may take a bit more doing. Naturally you must not tear the cardboard. Higher-order tetraflexagons of this type, if they have an even number of faces, can be constructed from similar rectangular starting patterns; tetraflexagons with an odd number of faces call for patterns analogous to the one used for the tri-tetraflexagon. Actually two rows of small squares are sufficient for making tetraflexagons of this sort, but adding one or more additional rows (which does not change the essential structure) makes the model easier to manipulate.

The tetra-tetraflexagon shown in Figure 9 has often been used as an advertising novelty because the difficulty of finding its fourth face makes it a pleasant puzzle. I have seen many such folders, some dating back to the 1930s. One had a penny glued to the hidden face; the object of the puzzle was to find the lucky penny. In 1946 Roger Montandon, of The Montandon Magic Company, Tulsa, Oklahoma, copyrighted a tetra-tetraflexagon folder called *Cherchez la Femme*, the puzzle being to find the picture of the young lady. Magic and novelty stores also sell an ancient children's trick usually called the "magic billfold." Its tri-tetraflexagon ribbon-hinges permit some simple disappearing stunts with a dollar bill and other flat objects.

A different variety of tetraflexagon, and one that has the unusual property of flexing along either of two axes at right angles to each

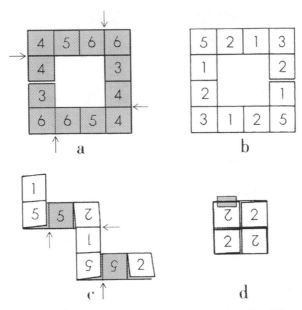

Figure 10. How to make a hexa-tetraflexagon. (Artist: Bunji Tagawa)

other, can also be made with four or more faces. The construction of a hexa-tetraflexagon of this type is depicted in Figure 10. Begin with the square-shaped strip shown in 10a (front) and 10b (back). Its small squares should be numbered as indicated. Crease along each internal line in 10a so that each line is the trough of a valley, flatten the strip again, and then fold on the four lines marked with arrows. All folds are made to conform with the way the lines were originally creased. The strip now looks like 10c. Fold on the three lines marked with arrows to form a square flexagon. Overlap the ends so that all the "2" squares are uppermost (10d). Attach a piece of transparent tape to the edge of the square at upper left, then bend it back to overlap the edge of a "1" square on the opposite side.

The hexa-tetraflexagon can now be flexed along both vertical and horizontal axes to expose all six of its faces. Larger square strips will yield flexagons whose number of faces increases by fours: 10, 14, 18, 22, and so on. For tetraflexagons of different orders, strips of other shapes must be used.

It was while Stone was working on right-triangle forms of flexagons ("for which, perhaps mercifully," he writes in a letter, "we invented no name") that he hit upon a most remarkable puzzle – the tetraflexatube. He had constructed a flat, square-shaped flexagon, which to his surprise opened into a tube. Further experimentation revealed that the tube could be turned completely inside out by a complicated series of flexes along the boundaries of the right triangles.

The flexatube is made from a strip of four squares (see Figure 11), each of which is ruled into four right triangles. Crease back and forth along all the lines, and then tape the ends together to form the cubical tube. The puzzle is to turn the tube inside out by folding only on the creased lines. A more durable version can be made by gluing 16 triangles of cardboard or thin metal onto cloth tape, allowing space between the triangles for flexing. It is useful to color only one side of the triangles, so that you can see at all times just what sort of progress you are making toward reversing the tube.

One method of solving this fascinating puzzle is illustrated in drawings 11b through 11k. Push the two A corners together, flattening the cube to the square flexagon of drawing 11c. Fold this forward along the axis BB to form the triangle of drawing 11d. Now push the two B corners together to make a flat square, but make sure that the two inside flaps go in opposite directions (11e). Open the square as in drawing 11f, then pull corner C down and to the left to make the flat structure shown in drawing 11g. Corner D is now pushed to the left, behind the structure, creating the flat rectangle of drawing 11h. This rectangle opens to form a cubical tube (11i) that is half the height of the original one.

You are now at the midpoint of your operations; exactly half the tube has been reversed. Flatten the tube to make a rectangle again (11j), but flatten it in the opposite way from that shown in drawing 11h. Starting as shown in drawing 11k, the previous operations are now "undone," so to speak, by performing them in reverse. Result: a reversed flexatube. At least two other completely different methods of turning the flexatube inside out are known, both as devious and difficult to discover as this one.

Recently Stone has been able to prove that a cylindrical band of *any* width can be turned inside out by a finite number of folds along straight lines, but the general method is much too involved to describe here. The question arises: Can a paper bag (that is, a

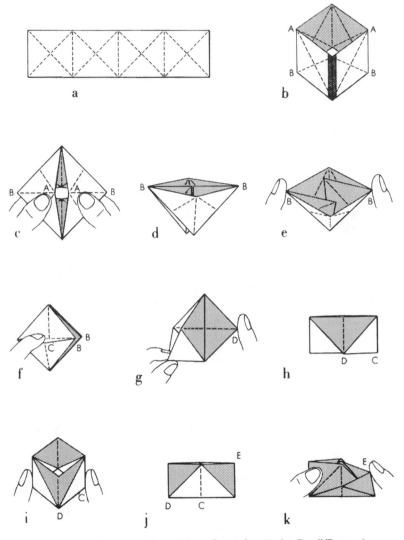

Figure 11. How to make and flex a flexatube. (Artist: Bunji Tagawa)

rectangular tube closed on the bottom) be turned inside out by a finite number of folds? This is an unsolved problem. Apparently the answer is no, regardless of the bag's proportions, though it probably would be extremely difficult to find a satisfactory proof.

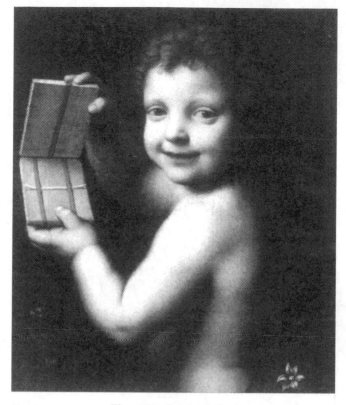

Figure 12. Luini painting.

POSTSCRIPT

Tetraflexagons, like hexaflexagons, are curious structures that belong to an obscure branch of mathematics called hinge theory. They permit a door, with hinges on both sides, to open to the left or to the right. Magic shops today sell what is called a Himber Wallet after Richard (Dick) Himber, an orchestra leader and the amateur magician who created it. It has two parts joined by a tetraflexagon hinge that allows it to be opened in two different ways to make cards appear and disappear. In 1998 Harry Lorayne published *The Himber Wallet Book* devoted entirely to tricks using the wallet.

It is hard to believe, but the wallet's structure is the same as that of a magic toy that goes back at least to 1520! That was the conjectured year that Bernardino Luini produced a beautiful oil painting of a small boy holding two blocks hinged by ribbons that make the blocks a tetraflexagon.

The cherub is causing a stick to transfer from one block to the other by opening the blocks in two different ways. It is the earliest known picture of a magic trick other than a painting of a street conjuror performing what magicians call a cups and balls routine.

The Luini painting (Figure 12) is reproduced in *Gibecièr* (Vol. 1, No. 1, Winter 2005), a handsome journal devoted to the worldwide history of conjuring. *Gibecièr* in turn took the picture from a British journal, *The Magic Circular* (January 1968), where it was wrongly attributed to Leonardo da Vinci.

The *Gibecièr* reproduction of Luini's painting accompanies an article by Volker Huber that traces the toy through many curious variants, notably the tumbling blocks. You'll find a picture of a *Cherchez la Femme* tetraflexagon on pages 361–363 of *Martin Gardner Presents* (1993), a book sold only in magic supply shops.

BIBLIOGRAPHY

"A Trick Book." "Willane" in *Willane's Wizardry*, pages 42–43. Privately printed in London, 1947. Shows how to construct the tetra-tetraflexagon depicted in Figure 9 of this book.

"A Deformation Puzzle." John Leech in *The Mathematical Gazette* 39:330 (December 1955): 307. The first printed description of the flexatube puzzle. No solution is given.

"Flexa Tube Puzzle." Martin Gardner in *Ibidem* 7 (a Canadian magic magazine) (September 1956): 13, with sample flexatube attached to page. A solution by T. S. Ransom appears in *Ibidem* 9 (March 1957): 12. Ransom's solution is the one given in this book.

Mathematical Snapshots. Hugo Steinhaus. Oxford University Press, revised edition, 1960. A series of photographs showing the solution of the flexatube puzzle that differs from Ransom's solution (see preceding entry) begins on page 190.

"Square Flexagons." P. B. Chapman in *Mathematical Gazette* 45 (1961): 193–194.

"Self-Designing Tetraflexagons." Robert E. Neale in *The Mathematician and Pied Puzzler*, eds. Elwyn Berlekamp and Tom Rodgers. A K Peters, 1999.

"It's Okay to Be Square if You're a Flexagon." Ethan J. Berkove and Jeffrey P. Dumont in *Mathematics Magazine* 17 (December 2004): 335–348.

CHAPTER THREE

Henry Ernest Dudeney: England's Greatest Puzzlist

HENRY ERNEST DUDENEY was England's greatest inventor of puzzles; indeed, he may well have been the greatest puzzlist who ever lived. Today there is scarcely a single puzzle book that does not contain (often without credit) dozens of brilliant mathematical problems that had their origin in Dudeney's fertile imagination.

He was born in the English village of Mayfield in 1857. Thus he was 16 years younger than Sam Loyd, the American puzzle genius. I do not know whether the two men ever met, but in the 1890s they collaborated on a series of puzzle articles for the English magazine *Tit-Bits*, and later they arranged to exchange puzzles for their magazine and newspaper columns. This may explain the large amount of duplication in the published writings of Loyd and Dudeney.

Of the two, Dudeney was probably the better mathematician. Loyd excelled in catching the public fancy with manufactured toys and advertising novelties. None of Dudeney's creations had the worldwide popularity of Loyd's "Get-off-the-Earth" paradox involving a vanishing Chinese warrior. On the other hand, Dudeney's work was mathematically more sophisticated (he once described the rebus or picture puzzle, of which Loyd produced hundreds, as a "juvenile imbecility" of interest only to the feeble-minded). Like Loyd, he enjoyed clothing his problems with amusing anecdotes. In this he may have had the assistance of his wife Alice, who wrote more than 30 romantic novels that were widely read in her time. His six books of puzzles (three are collections assembled after his death in 1930) remain unexcelled in the literature of puzzledom.

Dudeney's first book, *The Canterbury Puzzles*, was published in 1907. It purports to be a series of quaint posers propounded by the same group of pilgrims whose tales were recounted by Chaucer: "I

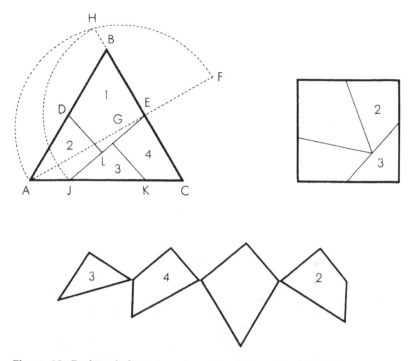

Figure 13. Dudeney's four-piece dissection of equilateral triangle to square. (Artist: Alex Semenoick)

will not stop to explain the singular manner in which they came into my possession," Dudeney writes, "but [will] proceed at once... to give my readers an opportunity of solving them." The haberdasher's problem, found in this book, is Dudeney's best-known geometrical discovery. The problem is to cut an equilateral triangle into four pieces that can then be reassembled to form a square.

The drawing at upper left in Figure 13 shows how the cuts are made. Bisect AB at D and BC at E. Extend AE to F so that EF equals EB. Bisect AF at G, then, with G as the center, describe the arc AHF. Extend EB to H. With E as the center, draw the arc HJ. Make JK equal to BE. From D and K drop perpendiculars on EJ to obtain points L and M. The four pieces can now be rearranged to make a perfect square, as shown at upper right in the illustration. A remarkable feature of this dissection is that, if the pieces are hinged at three vertices as shown in the drawing at the bottom, they form a chain that can be closed clockwise to make the triangle and counterclockwise to make the square. Dudeney rendered the figure into a brass-hinged mahogany model, which he used for demonstrating the problem before the Royal Society of London in 1905.

According to a theorem first proved by the great German mathematician David Hilbert, any polygon can be transformed into any other polygon of equal area by cutting it into a finite number of pieces. The proof is lengthy but not difficult. It rests on two facts: (1) any polygon can be cut by diagonals into a finite number of triangles, and (2) any triangle can be dissected into a finite number of parts that can be rearranged to form a rectangle of a given base. This means that we can change any polygon, however weird its shape, into a rectangle with a given base simply by chopping it first into triangles, changing the triangles to rectangles with the given base, and then piling the rectangles in a column. The column can then be used, by reversing the procedure, for producing any other polygon with an area equal to that of the original one.

Unexpectedly, the analogous theorem does not hold for polyhedrons: solids bounded by plane polygons. There is no general method for dissecting any polyhedron by plane cuts to form any different polyhedron of equal volume, though of course it can be done in special cases. Hope for a general method was abandoned in 1900 when it was proved impossible to dissect a prism into a regular tetrahedron.

Although Hilbert's procedure guarantees the transformation of one polygon into another by means of a finite number of cuts, the number of pieces required may be very large. To be elegant, a dissection must require the fewest possible pieces. This is often extremely difficult to determine. Dudeney was spectacularly successful in this odd geometrical art, often bettering long-established records. For example, although the regular hexagon can be cut into as few as five pieces that will make a square, the regular pentagon was for many years believed to require at least seven. Dudeney succeeded in reducing the number to six, the present record. Figure 14 shows how a pentagon can be squared by Dudeney's method. For an explanation of how Dudeney arrived at the method, the interested reader is referred to his *Amusements in Mathematics*, published in 1917.

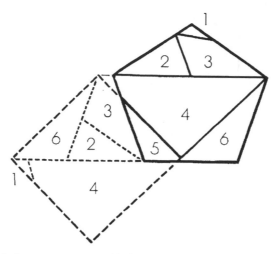

Figure 14. A pentagon reassembled into a square. (Artist: Alex Semenoick)

Dudeney's best-known brain teaser, about the spider and the fly, is an elementary but beautiful problem in geodesics. It first appeared in an English newspaper in 1903 but did not arouse widespread public interest until he presented it again two years later in the London *Daily Mail*. A rectangular room has the dimensions shown in Figure 15. The spider is at the middle of an end wall, one foot from the ceiling. The fly is at the middle of the opposite end wall, one foot above the floor, and too paralyzed with fear to move. What is the shortest distance the spider must crawl in order to reach the fly?

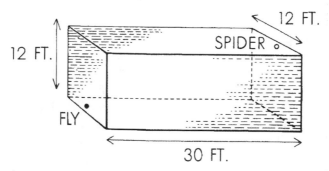

Figure 15. The problem of the spider and the fly. (Artist: Alex Semenoick)

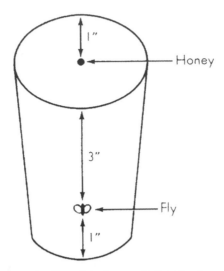

Figure 16. The fly and the honey. (Artist: Alex Semenoick)

The problem is solved by cutting the room so that walls and ceiling can be folded flat, then drawing a straight line from spider to fly. However, there are many ways in which the room can be folded flat, so it is not as easy as it first appears to determine the shortest path.

A less well-known but similar geodesic problem, which appears in Dudeney's *Modern Puzzles* (published in 1926), involves the cylindrical glass shown in Figure 16. It is four inches high and six inches in circumference. On the inside, one inch from the top, is a drop of honey. On the outside, one inch from the bottom and directly opposite, is a fly. What is the shortest path by which the fly can walk to the honey, and exactly how far does the fly walk?

It is interesting to note that although Dudeney had little familiarity with topology, then in its infancy, he frequently used clever topological tricks for solving various route and counter-moving puzzles. He called it his "buttons and string method." A typical example is afforded by the ancient chess problem shown in Figure 17. How can you make the white knights change places with the black knights in the fewest number of moves? We replace the eight outside squares with buttons (middle illustration) and draw lines to indicate all possible knight moves. If we regard these lines as strings joining the buttons, it is clear that we can open the string into a circle (bottom

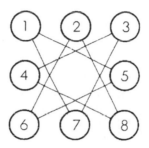

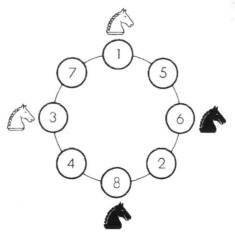

Figure 17. Dudeney's "buttons and string method." (Artist: Alex Semenoick)

illustration) without changing the topological structure of the elements and their connections. We see at once that we have only to move the knights around the circle in either direction until they are exchanged, keeping a record of the moves so that they can be reproduced on the original square board. In this way, what seems at first to be a difficult problem becomes ridiculously easy.

Of Dudeney's many problems involving number theory, perhaps the hardest to solve is the question posed by the doctor of physic in *The Canterbury Puzzles*. The good doctor produced two spherical phials, one exactly a foot in circumference and the other two feet in circumference. "I do wish," said the doctor, "to have the exact measures of two other phials, of a like shape but different in size, that may together contain just as much liquid as is contained by these two."

Since similar solids have volumes that are in the same proportion as the cubes of corresponding lengths, the problem reduces to the Diophantine task of finding two rational numbers other than 1 and 2 whose cubes will add up to nine. Both numbers must of course be fractions. Dudeney's solution was

$$\frac{415280564497}{348671682660} \quad \text{and} \quad \frac{676702467503}{348671682660}$$

These fractions had denominators of shorter length than any previously published. Considering the fact that Dudeney worked without a modern digital computer, the achievement is something to wonder at.

Readers who like this type of problem may enjoy the much simpler search for two fractions whose cubes total exactly six. A published "proof", by the nineteenth-century French mathematician Adrien Marie Legendre, that no such fractions could be found was exploded when Dudeney discovered a solution in which each fraction has only two digits above and two below the line.

ADDENDUM

Dudeney's dissection of the equilateral triangle to form a square brought a number of interesting letters from readers. John S. Gaskin of London and Arthur B. Niemoller of Morristown, New Jersey, independently discovered that Dudeney's method, with certain modifications, can be applied to a large class of triangles that are not equilateral. A lady in Brooklyn wrote that her son had constructed for her a nest of four tables, the tops of which can be fitted together to make either a square or an equilateral triangle, and that the tables had proved to be quite a conversation piece. L. Vosburgh Lyons of New York used Dudeney's construction for cutting the plane into an endless mosaic of interlocking squares and equilateral triangles.

Several readers, supposing that points J and K (in Figure 13) lay directly beneath points D and E, sent proofs that the four pieces would not form a perfect square. But Dudeney's construction does not put J and K exactly beneath D and E. A formal proof of the accuracy of the dissection will be found in Chester W. Hawley's article, "A Further Note on Dissecting a Square into an Equilateral Triangle," in *The Mathematics Teacher*, February 1960.

A remarkable variation of Dudeney's spider and fly problem will be found in Maurice Kraitchik's *Mathematical Recreations*, 1953, page 17. Eight spiders start from a spot 80 inches above the center of one end wall of the rectangular room. They take eight different paths to reach a fly that is 80 inches below the center of the opposite wall. Each spider moves at a speed of .65 mile per hour, and at the end of 625/11 seconds they arrive simultaneously at the fly. What are the room's dimensions?

ANSWERS

The shortest walking path of the spider to the fly is exactly 40 feet, as indicated on the unfolded room shown in Figure 18. The reader may be surprised that this geodesic carries the spider across five of the room's six sides.

The fly reaches the honey along the five-inch path drawn on the unrolled cylinder depicted in Figure 19. This is the path that would be taken by an imaginary beam of light moving across the rectangle from fly to honey and reflected by the rectangle's upper boundary. Clearly it is the same length as the hypotenuse of a right triangle with sides of three and four, as indicated.

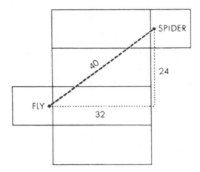

Figure 18. Answer to spider and fly problem. (Artist: Alex Semenoick)

The two fractions whose cubes add up to six are 17/21 and 37/21.

For an answer to the spiders and fly puzzle given in the addendum, consult the reference cited.

POSTSCRIPT

Greg Frederickson, the world's top expert on geometric dissections, has written an entire book titled *Piano-Hinged Dissections: Time to Fold!* (A K Peters, 2006). It is an amazing compilation of original discoveries of dissections in which pieces are hinged together so one polygon can be transformed to the other, like Dudeney's lovely triangle to square, simply by moving the hinged pieces.

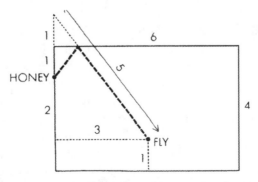

Figure 19. Answer to fly and honey problem. (Artist: Alex Semenoick)

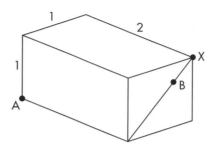

Figure 20. (Artist: Harold Jacobs)

Many puzzles have been based on an ant or a fly crawling over the surface of a specified solid. Consider the following problem. An ant is at corner A of a 1 × 1 × 2 brick. It crawls along a geodesic to point B somewhere on the 1 × 1 face opposite the starting face. Where should B be located to maximize the geodesic's length? Intuitively one would suppose it would be at corner X opposite A along a space diagonal because this is the point at the greatest distance from A. Not so! Yoshiyuki Kotani, a Japanese mathematician, recently discovered that the longest geodesic distance from A to B is to a point one-fourth of the way down a diagonal from the corner that is the farthest from A! (See Figure 20.)

For details and a proof see my article "The Ant on the 1 × 1 × 2" in *Math Horizons* (February 1996), reprinted in *Gardner's Workout* (A K Peters, 2001) and "Kotani's Ant Problem" by Dick Hess in *Puzzler's Tribute* (A K Peters, 2002). Hess discusses generalizations and variations of the problem. A section on "Spider and Fly Problems" is in David Singmaster's *Sources in Recreational Mathematics*, sixth edition (1993), published by the author and constantly updated on his Web site.

A difficult generalization of Dudeney's problem about the exchange of black and white knights on a 3 × 3 matrix can be found in Clifford Pickover's *A Passion for Mathematics* (Wiley, 2005), page 185. The matrix is 3 × 4. Four black knights, labeled A, B, C, and D, are along the top row. Four white knights are similarly labeled along the bottom row. The task is to use knight moves to exchange the two knights labeled A, the two labeled B, and so on for C and D, and to do this with the minimum number of moves. John Conway and Barry Cipra have proved that the smallest number of moves is 32.

I had the great pleasure of meeting Alice Dudeney, Henry's daughter, before she passed away. She told me that she did most of the illustrations for her father's puzzle books, and that her famous mother's diary was to be opened to the public in 2000. We arranged for Scribner's to combine Dudeney's last two books of mathematical puzzles into a single volume titled *536 Puzzles and Curious Problems* (1967). I regrouped the puzzles and wrote an introduction. The book was later reprinted by Barnes and Noble in 1995. Both editions are currently out of print.

In 1884 Dudeney married Alice Whiffin (1866–1945). She was then 18. According to the *Wikipedia,* Dudeney was a skillful pianist and organist, and a devout Anglican. He and Alice were for a time separated. Dudeney died of throat cancer in 1930. He and his wife are buried in Lewes, where they had moved in 1914.

Alice's personal diaries were edited by Diana Cook and published in 1998 under the title of *A Lewes Diary, 1916–1944.* "They give a lively picture," says the *Wikipedia,* "of her attempts to balance her literary career with her marriage to her brilliant but volatile husband."

BIBLIOGRAPHY

BOOKS BY DUDENEY

The Canterbury Puzzles, 1907. Reprinted by Dover Publications, Inc., in 1958.

Amusements in Mathematics, 1917. Reprinted by Dover Publications, Inc., in 1958.

Modern Puzzles, 1926.

Puzzles and Curious Problems, 1931.

A Puzzle-Mine, edited by James Travers, undated.

World's Best Word Puzzles, edited by James Travers, 1925.

ARTICLES BY DUDENEY

Puzzles and articles by Dudeney are scattered throughout many English newspapers and periodicals: *The Strand Magazine* (in which his puzzle column "Perplexities" ran for twenty years), *Cassell's Magazine*, *The Queen*, *The Weekly Dispatch*, *Tit-Bits*, *Educational Times*, *Blighty*, and others.

The following two articles are of special interest:

"The Psychology of Puzzle Crazes," in *The Nineteenth Century* 100:6 (December 1926): 868–879.

"Magic Squares," in *The Encyclopedia Britannica*, 14th ed.

REFERENCES ABOUT DUDENEY

"The Puzzle King: An Interview with Henry E. Dudeney." Fenn Sherie in *The Strand Magazine* 71 (April 1926): 398–404.

Preface by Alice Dudeney to her husband's *Puzzles and Curious Problems*, listed above.

"The Life and Work of H. E. Dudeney." Angela Newring in *Mathematical Spectrum* 21 (1988–1989): 38–44.

A biographical sketch of Alice Dudeney, who was more famous in her day than Henry, will be found in the British *Who Was Who*.

ON DISSECTIONS

Geometric Dissections. Harry Lindgren. D. Van Nostrand, 1964.

Dissections: Plane & Fancy. Greg N. Frederickson. Cambridge, 1997.

Hinged Dissections: Swinging & Twisting. Greg N. Frederickson. Cambridge, 2002.

Piano-Hinged Dissections. Greg N. Frederickson. A K Peters, 2006.

CHAPTER FOUR

Digital Roots

JOT DOWN your telephone number. Scramble the order of the digits in any way you please to form a new number; then subtract the smaller number from the larger. Add all the digits in the answer. Now place your finger on the star in the circle of mysterious symbols (Figure 21) and count them clockwise around the circle, calling the star 1, the triangle 2 and so on until you reach the number that was the final step in the procedure given above. Your count is sure to end on the spiral.

The operation of this little trick is not hard to understand, and it provides a painless introduction to the concept of numerical congruence formulated by the great German mathematician Carl Friedrich Gauss. If two numbers have the same remainder when divided by a given number called k, they are said to be congruent modulo k. The number k is called the modulus. Thus 16 and 23 both have a remainder of 2 when divided by 7 and are therefore congruent modulo 7.

Because 9 is the largest digit in the decimal number system, the sum of the digits of any number will always be congruent modulo 9 to the original number. The digits in this second number can then be added to obtain a third number congruent to the other two, and if we continue this process until only one digit remains, it will be the remainder itself. For example, 4,157 has a remainder of 8 when divided by 9. Its digits total 17, which also has a remainder of 8 modulo 9. And the digits of 17 add up to 8. This last digit is called the digital root of the original number. It is the same as the number's remainder modulo 9, with the exception of numbers with a remainder of 0, in which case the digital root is 9 instead of 0.

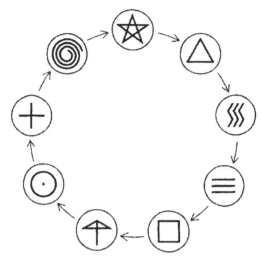

Figure 21. Symbols for a trick with a telephone number. (Artist: Alex Semenoick)

Obtaining the digital root is simply the ancient process of "casting out 9's." Before the development of computing devices, the technique was widely used by accountants for checking their results. Early electronic computers, for example, the International Business Machine NORC, used the technique as one of their built-in methods of self-checking for accuracy. The method is based on the fact that if whole numbers are added, subtracted, multiplied, or evenly divided, the answer will be congruent modulo 9 to the number obtaining by adding, subtracting, multiplying, or dividing the digital roots of those same numbers.

For example, to check quickly a sum involving large numbers you obtain the digital roots of the numbers, add them, reduce the answer to a root, then see if it corresponds to the digital root of the answer you wish to test. If the roots fail to match, you know that there is an error somewhere. If they do match, there still may be an error, but the probability is fairly high that the computation is correct.

Let us see how all this applies to the telephone-number trick. Scrambling the digits of the number cannot change its digital root, so we have here a case in which a number with a certain digital root is subtracted from a larger number with the same digital root. The result is certain to be a number evenly divisible by 9. To see why this is so, think of the larger number as consisting of a certain multiple of 9, to which is added a digital root (the remainder when the number is divided by nine). The smaller number consists of a smaller multiple of 9, to which is added the *same* digital root. When the smaller number is subtracted from the larger, the digital roots cancel out, leaving a multiple of 9.

$$
\begin{array}{l}
\text{(A multiple of 9)} + \text{a digital root} \\
-\text{(A multiple of 9)} + \text{the same digital root} \\
\hline
\text{(A multiple of 9)} + 0
\end{array}
$$

Since the answer is a multiple of 9, it will have a digital root of 9. Adding the digits will give a smaller number that also has a digital root of 9, so the final result is certain to be a multiple of 9. There are nine symbols in the mystic circle. The count, therefore, is sure to end on the ninth symbol from the first one that is tapped.

A knowledge of digital roots often furnishes amazing shortcuts in solving problems that seem unusually difficult. For example, suppose you are asked to find the smallest number composed of 1's and 0's that is evenly divisible by 225. The digits in 225 have a digital root of 9, so you know at once that the required number must also have a digital root of 9. The smallest number composed of 1's that will have a digital root of 9 is obviously 111,111,111. Adding zeros at significant spots will enlarge the number but will not alter the root. Our problem is to increase 111,111,111 by the smallest amount that will make it divisible by 225. Since 225 is a multiple of 25, the number we seek must also be a multiple of 25. All multiples of 25 must end in 00, 25, 50, or 75. The last three pairs cannot be used, so we attach 00 to 111,111,111 to obtain the answer: 11,111,111,100.

Mathematical games also frequently lend themselves to digital-root analysis, as for example this game played with a single die. An arbitrary number, usually larger than 20 to make the game interesting, is agreed upon. The first player rolls the die, scoring the number that is uppermost. The second player now gives the die a quarter turn in any direction, adding to the previous score the number he brings to the top. Players alternate in making quarter-turns, keeping a running total, until one of them wins by reaching the agreed-upon

number or forcing his opponent to go above it. The game is difficult
to analyze because the four side-numbers available at each turn
vary with the position of the die. What strategy should one adopt
to play the best possible game?

The key numbers in the strategy are those that have digital roots
that are the same as the digital root of the goal. If you can score a
number in this series, or permanently prevent your opponent from
doing so, you have a certain win. For example, the game is often
played with the goal of 31, which has a digital root of 4. The only
way the first player can force a win is by rolling a 4. Thereafter he
either plays to get back in the series 4–13–22–31, or plays so that his
opponent cannot enter it. Preventing an opponent from entering
the series is somewhat tricky, so I shall content myself with saying
only that one must either play to five below a key number, leaving
the 5 on the top or the bottom of the die; or to four or three below,
or one above, leaving the 4 on the top or the bottom.

There is always one roll, and sometimes two or three, that will
guarantee a win for the first player, except when the digital root of
the goal happens to be 9. In such cases, the second player can always
force a win. When the goal is chosen at random, the odds of winning
greatly favor the second player. If the first player chooses the goal,
what should be the digital root of the number he picks in order to
have the best chance of winning?

A large number of self-working card tricks depend on the properties of digital roots. In my opinion, the best is a trick currently sold
in magic shops as a four-page typescript titled "Remembering the
Future." It was invented by Stewart James of Courtright, Ontario, a
magician who has probably devised more high-quality mathematical card tricks than anyone who ever lived. The trick is explained
here with James's permission.

From a thoroughly shuffled deck you remove nine cards with values from ace to 9, arranging them in sequence with the ace on top.
Show the audience what you have done; then explain that you will
cut this packet so that no one will know what cards are at what positions. Hold the packet face down in your hands and appear to cut it
randomly but actually cut it so that three cards are transferred from
bottom to top. From the top down the cards will now be in the order:
7–8 9–1 2–3 4–5 6.

Slowly remove one card at a time from the top of this packet, transferring these cards to the top of the deck. As you take each card, ask a spectator if he wishes to select that card. He must, of course, select one of the nine. When he says "Yes," leave the chosen card on top of the remaining cards in the packet and put the packet aside.

The deck is now cut at any spot by a spectator to form two piles. Count the cards in one pile; then reduce this number to its digital root by adding the digits until a single digit remains. Do the same with the other pile. The two roots are now added, and if necessary the total is reduced to its digital root. The chosen card, on top of the packet placed aside, is now turned over. It has correctly predicted the outcome of the previous steps!

Why does it work? After the nine cards are properly arranged and cut, the 7 will be on top. The deck will consist of 43 cards, a number with a digital root of 7. If the spectator does not choose the 7, it is added to the deck, making a total of 44 cards. The packet now has an 8 on top, and 8 is the digital root of 44. In other words, the card selected by the spectator must necessarily correspond to the digital root of the number of cards in the deck. Cutting the deck in two parts and combining the roots of each portion as described will, of course, result in the same digit as the digital root of the entire deck.

ADDENDUM

It is asserted at the beginning of this chapter that because our number system is based on 10, the digital root of any number is the same as the remainder when that number is divided by 9. This is not hard to prove, and perhaps an informal statement of a proof will interest some readers.

Consider a four-digit number, say 4,135. This can be written as sums of powers of 10:

$$(4 \times 1{,}000) + (1 \times 100) + (3 \times 10) + (5 \times 1)$$

If 1 is subtracted from each power of 10, we can write the same number like this:

$$(4 \times 999) + (1 \times 99) + (3 \times 9) + (5 \times 0) + 4 + 1 + 3 + 5$$

The expressions inside the parentheses are all multiples of 9. After casting them out, we are left with 4 + 1 + 3 + 5, the digits of the original number.

In general, a number written with the digits *abcd* can be written:

$$(a \times 999) + (b \times 99) + (c \times 9) + (d \times 0) + a + b + c + d$$

Therefore $a + b + c + d$ must be a remainder after certain multiples of 9 are cast out. This remainder of course may be a number of more than one digit. If so, the same procedure will show that the sum of *its* digits will give another remainder after other multiples of 9 are cast out, and we can continue until only one digit, the digital root, remains. Such a procedure can be applied to any number, no matter how large. The digital root, therefore, is the number that remains after the maximum number of 9's have been cast out; that is, after the number is divided by 9.

Digital roots are often useful as negative checks in determining whether a very large number is a perfect square or cube. All square numbers have digital roots of 1, 4, 7, or 9, and the last digit of the number cannot be 2, 3, 7, or 8. A cube may end with any digit, but its digital root must be 1, 8, or 9. Most curiously of all, an even perfect number (and so far no odd perfect number has been found) must end in 6 or 28 and, with the exception of 6, the smallest perfect number, have a digital root of 1.

ANSWERS

In the game played with a die, if the first player chooses the number that is to be the goal, his best choice is a number with the digital root of 7. The chart in Figure 22 shows the winning first roll for each of the nine possible digital roots of the goal. Seven has three winning first rolls, more than any other digital root. This gives the first player a chance of 1/2 that he will roll a number that will lead to a sure win if he plays correctly.

POSTSCRIPT

Numerous puzzles and magic tricks based on digital roots are scattered throughout the books in this series. For now, I will add

DIGITAL ROOT OF GOAL	WINNING FIRST ROLLS OF DIE
1	1, 5
2	2, 3
3	3, 4
4	4
5	5
6	3, 6
7	2, 3, 4
8	4
9	NONE

Figure 22

only a riddle: In what country do 11 things plus 3 things, equal 2 things?

Answer: *Any* country. Eleven o'clock plus three more hours is two o'clock.

BIBLIOGRAPHY

"Doctor Daley's Thirty One." Jacob Daly in *The Conjuror's Magazine* (March and April 1945).

Remembering the Future. Stewart James. Sterling Magic Company, Royal Oak, Michigan, 1947.

"The Game of Thirty One." George G. Kaplan in *The Fine Art of Magic*, pages 275–279. Fleming Book Company, York, Pennsylvania, 1948.

"Magic with Pure Numbers." Martin Gardner in *Mathematics, Magic and Mystery.* Dover Publications, Inc., 1956.

CHAPTER FIVE

Nine Problems

1. THE TWIDDLED BOLTS

Two identical bolts are placed together so that their helical grooves intermesh (Figure 23). If you move the bolts around each other as you would twiddle your thumbs, holding each bolt firmly by the head so that it does not rotate and twiddling them in the direction shown, will the heads (a) move inward, (b) move outward, or (c) remain the same distance from each other? The problem should be solved without resorting to actual test.

2. THE FLIGHT AROUND THE WORLD

A group of airplanes is based on a small island. The tank of each plane holds just enough fuel to take it halfway around the world. Any desired amount of fuel can be transferred from the tank of one plane to the tank of another while the planes are in flight. The only source of fuel is on the island, and for the purposes of the problem it is assumed that there is no time lost in refueling either in the air or on the ground.

Figure 23. The twiddled bolts. (Photo: Peter Reuz)

What is the smallest number of planes that will ensure the flight of one plane around the world on a great circle, assuming that the planes have the same constant ground speed and rate of fuel consumption and that all planes return safely to their island base?

3. THE CIRCLE ON THE CHESSBOARD

A chessboard has squares that are two inches on the side. What is the radius of the largest circle that can be drawn on the board in such a way that the circle's circumference is entirely on black squares?

4. THE CORK PLUG

Many old puzzle books explain how a cork can be carved to fit snugly into square, circular and triangular holes (Figure 24). An interesting problem is to find the volume of the cork plug. Assume that it has a circular base with a radius of one unit, a height of two units, and a straight top edge of two units that is directly above and parallel to a diameter of the base. The surface is such that all vertical cross sections made perpendicular to the top edge are triangles.

The surface may also be thought of as generated by a straight line connecting the sharp edge with the circular edge and moving so that it is at all times parallel to a plane that is perpendicular to the sharp edge. The plug's volume can of course be determined by calculus, but there is a simple way to find it with little more information than knowing that the volume of a right circular cylinder is the area of its base times its altitude.

5. THE REPETITIOUS NUMBER

An unusual parlor trick is performed as follows. Ask spectator A to jot down any three-digit number, and then to repeat the digits in the same order to make a six-digit number (e.g., 394,394). With your back turned so that you cannot see the number, ask A to pass the sheet of paper to spectator B, who is requested to divide the number by 7.

"Don't worry about the remainder," you tell him, "because there won't be any." B is surprised to discover that you are right (e.g., 394,394 divided by 7 is 56,342). Without telling you the result, he

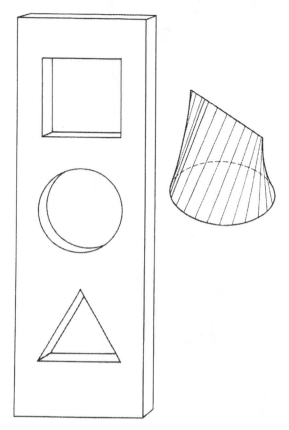

Figure 24. The cork plug.

passes it on to spectator C, who is told to divide it by 11. Once again you state that there will be no remainder, and this also proves correct (56,342 divided by 11 is 5,122).

With your back still turned, and no knowledge whatever of the figures obtained by these computations, you direct a fourth spectator, D, to divide the last result by 13. Again the division comes out even (5,122 divided by 13 is 394). This final result is written on a slip of paper which is folded and handed to you. Without opening it you pass it on to spectator A.

"Open this," you tell him, "and you will find your original three-digit number."

Prove that the trick cannot fail to work regardless of the digits chosen by the first spectator.

Figure 25. The sliding pennies (Artist: David Jou).

6. THE COLLIDING MISSILES

Two missiles speed directly toward each other, one at 9,000 miles per hour and the other at 21,000 miles per hour. They start 1,317 miles apart. Without using pencil and paper, calculate how far apart they are one minute before they collide.

7. THE SLIDING PENNIES

Six pennies are arranged on a flat surface as shown in Figure 25. The problem is to move them into the formation depicted at bottom in the smallest number of moves. Each move consists in sliding a penny, without disturbing any of the other pennies, to a new position in which it touches two others. The coins must remain flat on the surface at all times.

8. HANDSHAKES AND NETWORKS

Prove that at a recent convention of biophysicists the number of scientists in attendance who shook hands an odd number of times is even. The same problem can be expressed graphically as follows. Put as many dots (biophysicists) as you wish on a sheet of paper. Draw as many lines (handshakes) as you wish from any dot to any other dot. A dot can "shake hands" as often as you please, or not at all. Prove that the number of dots with an odd number of lines joining them is even.

9. THE TRIANGULAR DUEL

Smith, Brown, and Jones agree to fight a pistol duel under the following unusual conditions. After drawing lots to determine who fires first, second, and third, they take their places at the corners of an equilateral triangle. It is agreed that they will fire single shots in turn and continue in the same cyclic order until two of them are dead. At each turn the man who is firing may aim wherever he pleases. All three duelists know that Smith always hits his target, Brown is 80 percent accurate, and Jones is 50 percent accurate.

Assuming that all three adopt the best strategy, and that no one is killed by a wild shot not intended for him, who has the best chance to survive? A more difficult question: What are the exact survival probabilities of the three men?

ANSWERS

1. The heads of the twiddled bolts move neither inward nor outward. The situation is comparable to that of a person walking up an escalator at the same rate that it is moving down. (I am indebted to Theodore A. Kalin for calling this problem to my attention.)

2. Three airplanes are quite sufficient to ensure the flight of one plane around the world. There are many ways this can be done, but the following seems to be the most efficient. It uses only five

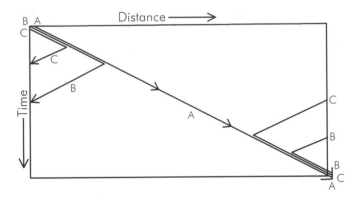

Figure 26. The flight around the world.

tanks of fuel, allows the pilots of two planes sufficient time for a cup of coffee and a sandwich before refueling at the base, and there is a pleasing symmetry in the procedure.

Planes A, B, and C take off together. After going 1/8 of the distance, C transfers 1/4 tank to A and 1/4 to B. This leaves C with 1/4 tank; just enough to get back home.

Planes A and B continue another 1/8 of the way, then B transfers 1/4 tank to A. B now has 1/2 tank left, which is sufficient to get him back to the base where he arrives with an empty tank.

Plane A, with a full tank, continues until it runs out of fuel 1/4 of the way from the base. It is met by C which has been refueled at the base. C transfers 1/4 tank to A, and both planes head for home.

The two planes run out of fuel 1/8 of the way from the base, where they are met by refueled plane B. Plane B transfers 1/4 tank to each of the other two planes. The three planes now have just enough fuel to reach the base with empty tanks.

The entire procedure can be diagrammed as shown in Figure 26, where distance is the horizontal axis and time the vertical axis. The right and left edges of the chart should, of course, be regarded as joined.

3. If you place the point of a compass at the center of a black square on a chessboard with two-inch squares, and extend the arms of the compass a distance equal to the square root of 10 inches, the pencil will trace the largest possible circle that touches only black squares.

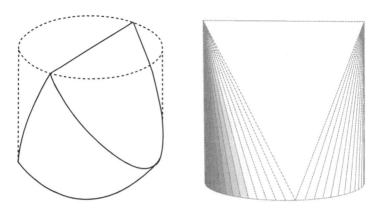

Figure 27. Slicing the cork, minimal convex plug.

4. Any vertical cross section of the cork plug at right angles to the top edge and perpendicular to the base will be a triangle. If the cork were a cylinder of the same height, corresponding cross sections would be rectangles. Each triangular cross section is obviously 1/2 the area of the corresponding rectangular cross section. Since all the triangular sections combine to make up the cylinder, the plug must be 1/2 the volume of the cylinder. The cylinder's volume is 2π, so our answer is simply π. (This solution is given in "No Calculus, Please," by J. H. Butchart and Leo Moser in *Scripta Mathematica*, September–December 1952.)

Actually, the cork can have an infinite number of shapes and still fit the three holes. The largest volume is obtained by the simple procedure of slicing the cylinder with two plane cuts as shown in Figure 27. This is the shape given in most puzzle books that include the plug problem. Its volume is equal to $2\pi - 8/3$. (I am indebted to J. S. Robertson, East Setauket, New York, for sending this calculation.) The minimal convex plug is also shown in Figure 27; the volume is $1/3 (2\pi + 4)$.

5. Writing a three-digit number twice is the same as multiplying it by 1,001. This number has the factors 7, 11, and 13, so writing the chosen number twice is equivalent to multiplying it by 7, 11, and 13. Naturally when the product is successively divided by these same three numbers, the final remainder will be the original number.

6. The two missiles approach each other with combined speeds of 30,000 miles per hour, or 500 miles per minute. By running the scene backward in time, we see that one minute before the collision the missiles would have to be 500 miles apart.

7. Number the top coin in the pyramid 1, the coins in the next row 2 and 3, and those in the bottom row 4, 5, and 6. The following four moves are typical of many possible solutions: Move 1 to touch 2 and 4, move 4 to touch 5 and 6, move 5 to touch 1 and 2 below, move 1 to touch 4 and 5.

8. Because two people are involved in every handshake, the total score for everyone at the convention will be evenly divisible by two and therefore even. The total score for the men who shook hands an even number of times is, of course, also even. If we subtract this even score from the even total score of the convention, we get an even total score for those men who shook hands an odd number of times. Only an even number of odd numbers will total an even number, so we conclude that an even number of men shook hands an odd number of times.

    There are other ways to prove the theorem; one of the best was sent to me by Gerald K. Schoenfeld, a medical officer in the U.S. Navy. At the start of the convention, before any handshakes have occurred, the number of persons who have shaken hands an odd number of times will be 0. The first handshake produces two "odd persons." From now on, handshakes are of three types: between two even persons, two odd persons, or one odd and one even person. Each even-even shake increases the number of odd persons by 2. Each odd-odd shake decreases the number of odd persons by 2. Each odd-even shake changes an odd person to even and an even person to odd, leaving the number of odd persons unchanged. There is no way, therefore, that the even number of odd persons can shift its parity; it must remain at all times an even number.

    Both proofs apply to a graph of dots on which lines are drawn to connect pairs of dots. The lines form a network on which the number of dots that mark the meeting of an odd number of lines is even. This theorem will be encountered again in Chapter 7 in connection with network-tracing puzzles.

9. In the triangular pistol duel the poorest shot, Jones, has the best chance to survive. Smith, who never misses, has the second best

chance. Because Jones's two opponents will aim at each other when their turns come, Jones's best strategy is to fire into the air until one opponent is dead. He will then get the first shot at the survivor, which gives him a strong advantage.

Smith's survival chances are the easiest to determine. There is a chance of 1/2 that he will get the first shot in his duel with Brown, in which case he kills him. There is a chance of 1/2 that Brown will shoot first, and since Brown is 4/5 accurate, Smith has a 1/5 chance of surviving. So Smith's chance of surviving Brown is 1/2 added to $1/2 \times 1/5 = 3/5$. Jones, who is accurate half the time, now gets a crack at Smith. If he misses, Smith kills him, so Smith has a survival chance of 1/2 against Jones. Smith's overall chance of surviving is therefore $3/5 \times 1/2 = 3/10$.

Brown's case is more complicated because we run into an infinite series of possibilities. His chance of surviving against Smith is 2/5 (we saw earlier that Smith's survival chance against Brown was 3/5, and since one of the two men must survive, we subtract 3/5 from 1 to obtain Brown's chance of surviving against Smith). Brown now faces fire from Jones. There is a chance of 1/2 that Jones will miss, in which case Brown has a 4/5 chance of killing Jones. Up to this point, then, his chance of killing Jones is $1/2 \times 4/5 = 4/10$. But there is a 1/5 chance that Brown may miss, giving Jones another shot. Brown's chance of surviving is 1/2; then he has a 4/5 chance of killing Jones again, so his chance of surviving on the second round is $1/2 \times 1/5 \times 1/2 \times 4/5 = 4/100$.

If Brown misses again, his chance of killing Jones on the third round will be 4/1,000; if he misses once more, his chance on the fourth round will be 4/10,000, and so on. Brown's total survival chance against Jones is thus the sum of the infinite series:

$$\frac{4}{10} + \frac{4}{100} + \frac{4}{1,000} + \frac{4}{10,000} \cdots$$

This can be written as the repeating decimal 0.444444..., which is the decimal expansion of 4/9.

As we saw earlier, Brown had a 2/5 chance of surviving Smith; now we see that he has a 4/9 chance to survive Jones. His overall survival chance is therefore $2/5 \times 4/9 = 8/45$.

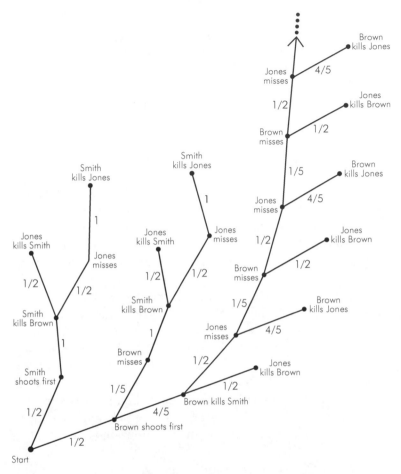

Figure 28. A tree diagram of the pistol-duel problem.

Jones's survival chance can be determined in similar fashion, but of course we can get it immediately by subtracting Smith's chance, 3/10, and Brown's chance, 8/45, from 1. This gives Jones an overall survival chance of 47/90.

The entire duel can be conveniently graphed by using the tree diagram shown in Figure 28. It begins with only two branches because Jones passes if he has the first shot, leaving only two equal possibilities: Smith shooting first or Brown shooting first, with intent to kill. One branch of the tree goes on endlessly.

The overall survival chance of an individual is computed as follows:

1. Mark all the ends of branches at which the person is sole survivor.
2. Trace each end back to the base of the tree, multiplying the probabilities of each segment as you go along. The product will be the probability of the event at the end of the branch.
3. Add together the probabilities of all the marked endpoint events. The sum will be the overall survival probability for that person.

In computing the survival chances of Brown and Jones, an infinite number of endpoints are involved, but it is not difficult to see from the diagram how to formulate the infinite series that is involved in each case.

When I published the answer to this problem I added that perhaps there is a moral of international politics in this somewhere. This prompted the following comment from Lee Kean of Dayton, Ohio:

Sirs:
  We must not expect that in international politics nations will behave as sensibly as individuals. Fifty-fifty Jones, against his own best interests, will blaze away when able at the opponent he imagines to be most dangerous. Even so, he still has the best chance of survival, 44.722 per cent. Brown and Smith find their chances reversed. Eighty-twenty Brown's chances are 31.111 per cent and sure-shot Smith comes in last with 24.167 per cent. Maybe the moral for international politics is even better here.

The problem, in variant forms, appears in several puzzle books. The earliest reference known to me is Hubert Phillips's *Question Time*, 1938, Problem 223. A different version of the problem can be found in Clark Kinnaird's *Encyclopedia of Puzzles and Pastimes*, 1946, but the answer is incorrect. Correct probability figures for Kinnaird's version are given in *The American Mathematical Monthly*, December 1948, page 640.

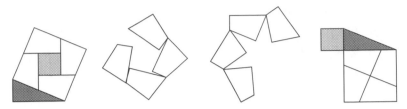

Figure 29. (Artist: Harold Jacobs)

POSTSCRIPT

Gwen Roberts, writing on "Shadows and Plugs" in *Puzzler's Tribute* (edited by David Wolfe and Tom Rodgers, A K Peters, 2002), introduces a variant of the three holes and one plug. The holes are a circle, cross, and square. Gwen's high school students discovered that the three solids forming the traditional plug – cone, sphere, and cylinder – have volumes in 1:2:3 ratios.

Ms. Roberts included a picture of a beautiful hinged-blocks proof of the Pythagorean Theorem (see Figure 29).

David Singmaster, writing "On Round Pegs in Square Holes, and Square Pegs in Round Holes," *Mathematics Magazine* (Vol. 37, 1964, pages 335–337), concludes, using the ratio of areas, that a round peg fits more snugly in a square hole than a square peg in a round hole. The ratios are $\pi/4$ which is greater than $2/\pi$. A generalization to n dimensions reveals the astonishing fact that the n-ball fits better in the n-cube than the n-cube fits in the n-ball, if and only if n is equal to or less than 8.

There is now a growing literature on triangular duels. Donald Knuth, in 1973, wrote a paper titled "The Triel: A New Solution," in which he reasoned that the optimum strategy for all three duelists is to fire in the air! See *The Journal of Recreational Mathematics*, Vol. 6 (pp. 1–17). His controversial solution applies only to three players. There are current Web sites devoted to the general problem and a *Wikipedia* article.

CHAPTER SIX

The Soma Cube

[N]o time, no leisure . . . not a moment to sit down and think – of if ever by some unlucky chance such a crevice of time should yawn in the solid substance of their distractions, there is always *soma*, delicious *soma*.

Aldous Huxley, *Brave New World*

THE CHINESE PUZZLE GAME called tangrams employs a square of thin material that is dissected into seven pieces (see Chapter 18). The game is to rearrange those pieces to form other figures. From time to time efforts have been made to devise a suitable analog in three dimensions. None, in my opinion, has been as successful as the Soma cube, invented by Piet Hein, the Danish writer whose mathematical games, Hex and Tac Tix, are discussed in my Book 1. (In Denmark, Piet Hein was best known for his books of epigrammatic poems written under the pseudonym Kumbel.)

Piet Hein conceived of the Soma cube during a lecture on quantum physics by Werner Heisenberg. While the noted German physicist was speaking of a space sliced into cubes, Piet Hein's supple imagination caught a fleeting glimpse of the following curious geometrical theorem. If you take all the irregular shapes that can be formed by combining no more than four cubes, all the same size and joined at their faces, these shapes can be put together to form a larger cube.

Let us make this clearer. The simplest irregular shape – "irregular" in the sense that it has a concavity or corner nook in it somewhere – is produced by joining three cubes as shown in Figure 30, piece 1. It is the only such shape possible with three cubes. (Of course no

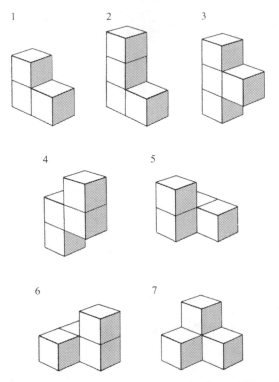

Figure 30. The seven Soma pieces. (Artist: James Egleson)

irregular shape is possible with one or two cubes.) Turning to four
cubes, we find that there are six different ways to form irregular
shapes by joining the cubes face to face. These are pieces 2 to 7 in
the illustration. To identify the seven pieces Piet Hein labels them
with numerals. No two shapes are alike, although 5 and 6 are mirror images of each other. Piet Hein points out that two cubes can be
joined only along a single coordinate, three cubes can add a second
coordinate perpendicular to the first, and four cubes are necessary
to supply the third coordinate perpendicular to the other two. Since
we cannot enter the fourth dimension to join cubes along a fourth
coordinate supplied by five-cube shapes, it is reasonable to limit our
set of Soma pieces to seven. It is an unexpected fact that these elementary combinations of identical cubes can be joined to form a
cube again.

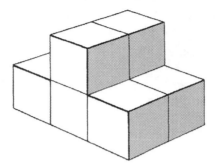

Figure 31. A form made up of two Soma pieces. (Artist: James Egleson)

As Heisenberg talked on, Piet Hein swiftly convinced himself by doodling on a sheet of paper that the seven pieces, containing 27 small cubes, would form a 3 × 3 × 3 cube. After the lecture he glued 27 cubes into the shapes of the seven components and quickly confirmed his insight. A set of the pieces was marketed under the trade name Soma, and the puzzle has since become a popular one in the Scandinavian countries.

To make a Soma cube – and the reader is urged to do so, for it provides a game that will keep every member of the family entranced for hours – you have only to obtain a supply of children's blocks. The seven pieces are easily constructed by spreading rubber cement on the appropriate faces, letting them dry, and then sticking them together.

As a first lesson in the art of Soma, see if you can combine any two pieces to form the stepped structure in Figure 31. Having mastered this trivial problem, try assembling all seven pieces into a cube. It is one of the easiest of all Soma constructions. More than 230 essentially different solutions (not counting rotations and reflections) have been tabulated by Richard K. Guy of the University of Malaya in Singapore, but the exact number of such solutions has not yet been determined. A good strategy to adopt on this as well as other Soma figures is to set the more irregular shapes (5, 6, and 7) in place first, because the other pieces adjust more readily to remaining gaps in a structure. Piece 1 in particular is best saved until last.

After solving the cube, try your hand at the more difficult seven-piece structures in Figure 32. Instead of using a time-consuming

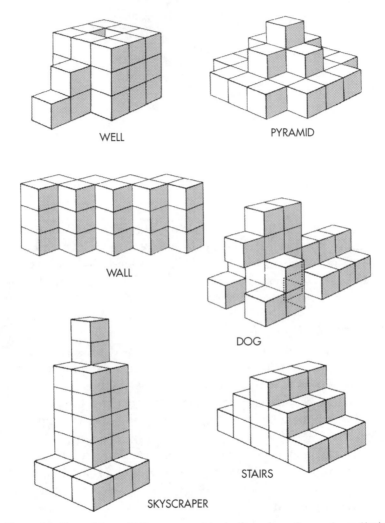

WELL

PYRAMID

WALL

DOG

SKYSCRAPER

STAIRS

Figure 32. One of these 12 forms cannot be built up from Soma pieces (Artist: James Egleson)

trial-and-error technique, it is much more satisfying to analyze the constructions and cut down your building time by geometrical insights. For example, it is obvious that pieces 5, 6, and 7 cannot form the steps to the well. Group competition can be introduced by giving each player a Soma set and seeing who can build

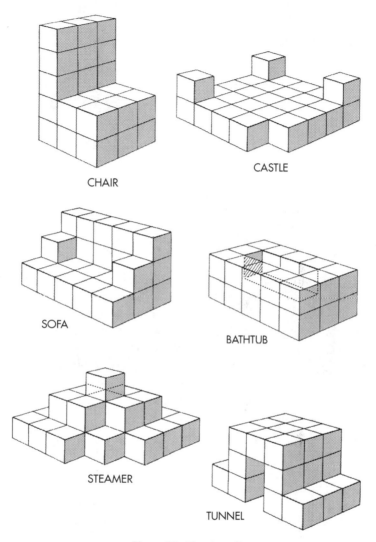

CHAIR

CASTLE

SOFA

BATHTUB

STEAMER

TUNNEL

Figure 32 (Continued)

a given figure in the shortest length of time. To avoid misinterpretations of these structures it should be said that the far sides of the pyramid and steamer are exactly like the near sides; both the hole in the well and the interior of the bathtub have a volume of three

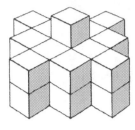

An impossible Soma form.   A means of labeling the form.

Figure 33. (Artist: James Egleson)

cubes; there are no holes or projecting pieces on the hidden sides of the skyscraper; and the column that forms the back of the dog's head consists of four cubes, the bottom one of which is hidden from view.

After working with the pieces for several days, many people find that the shapes become so familiar that they can solve Soma problems in their heads. Tests made by European psychologists have shown that ability to solve Soma problems is roughly correlated with general intelligence, but with peculiar discrepancies at both ends of the I. Q. curve. Some geniuses are very poor at Soma and some morons seem specially gifted with the kind of spatial imagination that Soma exercises. Everyone who takes such a test wants to keep playing with the pieces after the test is over.

Like the two-dimensional polyominoes, Soma constructions lend themselves to fascinating theorems and impossibility proofs of combinatorial geometry. Consider the structure in the left illustration of Figure 33. No one had succeeded in building it, and eventually a formal impossibility proof was devised. Here is the clever proof, discovered by Solomon W. Golomb, mathematician at the University of Southern California.

We begin by looking down on the structure as shown in the right illustration and coloring the columns in checkerboard fashion. Each column is two cubes deep except for the center column, which consists of three cubes. This gives us a total of eight white cubes and 19 black, quite an astounding disparity.

The next step is to examine each of the seven components, testing it in all possible orientations to ascertain the maximum number

SOMA PIECE	MAXIMUM BLACK CUBES	MINIMUM WHITE CUBES
1.	2	1
2.	3	1
3.	3	1
4.	2	2
5.	3	1
6.	3	1
7.	2	2
	18	9

Figure 34. Table for the impossibility proof.

of black cubes it can possess if placed within the checkerboard structure. The chart in Figure 34 displays this maximum number for each piece. As you see, the total is 18 black to nine white, just one short of the 19–8 split demanded. If we shift the top black block to the top of one of the columns of white blocks, then the black–white ratio changes to the required 19/8, and the structure becomes possible to build.

I must confess that one of the structures in Figure 32 is impossible to make. It should take the average reader many days, however, to discover which one it is. Methods for building the other figures will not be given in the answer section (it is only a matter of time until you succeed with any one of them), but I shall identify the figure that cannot be made.

The number of pleasing structures that can be built with the seven Soma pieces seems to be as unlimited as the number of plane figures that can be made with the seven tangram shapes. It is interesting to note that if piece 1 is put aside, the remaining six pieces will form a shape exactly like 1 but twice as high.

ADDENDUM

When I wrote the column about Soma, I supposed that few readers would go to the trouble of actually making a set. I was wrong.

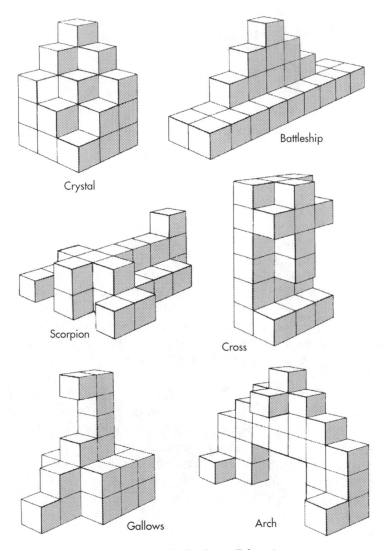

Figure 35. (Artist: James Egleson)

Thousands of readers sent sketches of new Soma figures and many complained that their leisure time had been obliterated since they were bitten by the Soma bug. Teachers made Soma sets for their classes. Psychologists added Soma to their psychological tests. Somaddicts made sets for friends in hospitals and gave them as

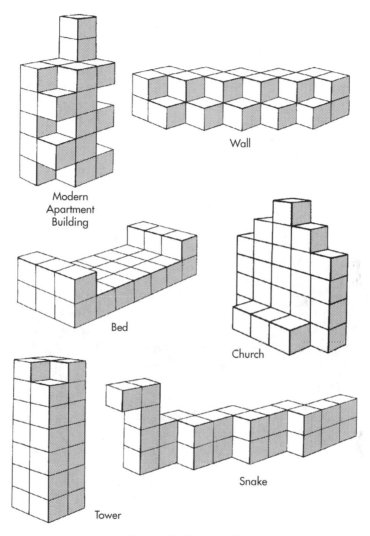

Figure 35 (Continued)

Christmas gifts. A dozen firms inquired about manufacturing rights. From the hundreds of new Soma figures received from readers, I have selected the twelve that appear in Figure 35. Some of these figures were discovered by more than one reader. All are possible to construct.

The charm of Soma derives in part, I think, from the fact that only seven pieces are used; one is not overwhelmed by complexity. All sorts of variant sets, with a larger number of pieces, suggest themselves, and I have received many letters describing them.

Theodore Katsanis of Seattle, in a letter dated December 23, 1957 (before the article on Soma appeared), proposed a set consisting of the eight different pieces that can be formed with four cubes. This set includes six of the Soma pieces plus a straight chain of four cubes and a 2×2 square. Katsanis called them "quadracubes"; other readers later suggested "tetracubes." The eight pieces will not, of course, form a cube; but they do fit neatly together to make a $2 \times 4 \times 4$ rectangular solid. This is a model, twice as high, of the square tetracube. It is possible to form similar models of each of the other seven pieces. Katsanis also found that the eight pieces can be divided into two sets of four, each set making a $2 \times 2 \times 4$ rectangular solid. These two solids can then be put together in different ways to make double-sized models of six of the eight pieces.

In a previous column (Chapter 13 of Book 1), I described the 12 pentominoes: flat shapes formed by connecting unit squares in all possible ways. Mrs. R. M. Robinson, wife of a mathematics professor at the University of California at Berkeley, discovered that if the pentominoes are given a third dimension, one unit thick, the 12 pieces will form a $3 \times 4 \times 5$ rectangular solid. This was independently discovered by several others, including Charles W. Stephenson, M.D., of South Hero, VT. Dr. Stephenson also found ways of putting together the three-dimensional pentominoes to make rectangular solids of $2 \times 5 \times 6$ and $2 \times 3 \times 10$.

The next step in complexity is to the 29 pieces formed by putting five cubes together in all possible ways. Katsanis, in the same letter mentioned previously, suggested this and called the pieces "pentacubes." Six pairs of pentacubes are mirror-image forms. If we use only one of each pair, the number of pentacubes drops to 23. Both 29 and 23 are primes; therefore, no rectangular solids are possible with either set. Katsanis proposed a triplication problem: choose one of the 29 pieces, then use 27 of the remaining 28 to form a model of the selected piece, three times as high.

A handsome set of pentacubes was shipped to me in 1960 by David Klarner of Napa, CA. I dumped them out of the wooden box in which they were packed and have not yet succeeded in putting them back in. Klarner has spent considerable time developing unusual pentacube figures, and I have spent considerable time trying to build some of them. He writes that there are 166 *hexacubes* (pieces formed by joining six-unit cubes), of which he was kind enough *not* to send a set.

The seven Soma pieces are a subset of what are now called polycubes – polyhedrons formed by joining unit cubes by their faces. Since I introduced the Soma cube in a 1958 column it has been made and sold by numerous toy companies around the world. Here Parker Brothers sold the cube along with an instruction booklet written by Piet Hein. The firm also distributed three issues of *Soma Addict*, a newsletter edited by game agent Thomas Atwater.

Several computer programs verified that there are 240 ways, not counting rotations and reflections, to make the Soma cube. John Conway produced what he called the Somap. You'll find a picture of it in *Winning Ways*, Vol. 4, by Elwyn Berlekamp, Conway, and Richard Guy, pages 911–912. This amazing graph shows how you can start with any of 239 solutions to the cube and then transform it to any other solution by moving no more than two or three pieces. There is one solution unobtainable in this way.

J. Edward Hanrahan wrote to me about a Soma task he invented. The challenge is to form a $4 \times 4 \times 2$ structure so that its five "holes" on the top layer have the shape of each of the 12 pentominoes. The problem is solvable for each pentomino except the straight one, which obviously can't fit into the structure.

Puzzle collector Jerry Slocum owns dozens of different dissections of a 3^3 cube into seven or fewer pieces, most of them marketed after my column on the Soma cube appeared. In the chapter of polycubes in Book 11, I describe the Diabolical cube, sold in Victorian England. The earliest dissection of a 3^3 cube known, its six polycube pieces form the cube in 13 ways. I also describe the Miksinski cube, another six-piece dissection – one that has only two solutions.

Many later 3^3 dissections limit the number of solutions by coloring or decorating the unit cube faces in various ways. For example, the unit cubes are either black or white, and the task is to make a cube that is checkered throughout or just on its six faces. Other dissections put one through six spots on the unit cubes so that the assembled cube will resemble a die. The unit cubes can be given different colors. The task is to form a cube with a specified pattern of colors on each face. Another marketed cube had digits 1 through 9 on the unit cubes, and the problem was to make a cube with each face a magic square. A puzzle company in Madison, WI, advertised a game using a set of nine polycubes. Players selected any seven at random by rolling a pair of dice, and then tried to make a 3^3 cube with the pieces. In 1969, six different dissections of the cube, each made with five, six, or seven polycubes, were sold under the name Impuzzibles.

It's easy to cut a 3^3 cube into six or fewer polycubes that will make a cube in just one way, but with seven unmarked pieces, as I said earlier, it is not so easy. Rhoma, a slant version of Soma produced by a shear distortion that changes each unit cube as well as the large cube to a rhomboid shape, went on sale here. A more radically squashed Soma was sold in Japan. With such distortions the solution becomes unique.

John Brewer, of Lawrence, KS, in a little magazine he used to publish called *Hedge Apple and Devil's Claw* (Autumn 1995 issue), introduced the useful device of giving each Soma piece a different color. Solutions could then be represented by showing three sides of the cube with its unit cubes properly colored. He sent me a complete Somap using such pictures for each solution. His article also tells of his failed effort to locate Marguerite Wilson, the first to publish a complete set of solutions to the Soma cube.

Alan Guth, the M.I.T. physicist famous for his conjecture that a moment after the big bang the universe rapidly inflated, was quoted as follows in *Discover* (December 1997):

> My all-time favorite puzzle was a game called Soma, which I think was first marketed when I was in college. A set consisted of seven odd-shaped pieces that could be put together to make a cube, or a large variety of other three-dimensional shapes. Playing with two sets at once was even more fun. The instruction book claims that there are a certain number of different ways of making the cube, and I remember writing a computer program to verify this number. They were right, but I recall that they counted each of the 24 different orientations of the cube as a different "way" of making it.

Figure 36. (Photo: Ed Vogel)

POSTSCRIPT

In September 2006 the Minnesota State Fair, in St. Paul, featured on its grounds a huge set of Soma cube pieces constructed by Ed Vogel, of Minneapolis. He sent me two CDs, one containing many color photos of the exhibit, the other showing stages during its construction. One of the photographs is reproduced here with Vogel's kind permission (Figure 36).

Vogel writes that he became addicted to Soma after reading this chapter when it first appeared as a *Scientific American* column. Years later he constructed several large Soma cubes from cardboard boxes, culminating in the world's largest Soma, which he built in 2006 with the help of his friend Steven Jevning. When I asked how he wished to be identified, he replied in a letter, "proud jack of all trades and admitted master of none."

More on the Soma cube can be found in the chapter on polycubes in my Book 11.

Donald Knuth, going through my files in 2007, found a letter from Anneke Treep, dated late in 1988, in which she described seven

A

B

C

Figure 37. (Photo: Peter Renz)

polycubes which she claimed would form a cube in only one way. The pieces are shown in Figure 37. I had then no way to confirm or refute her claim. Knuth found it a simple matter to write a computer program that proved Anneke was correct. She later found several other solutions of a similar form. Still another, also similar, was sent to me by Peter van den Muijzenberg.

A completely different solution, using seven pieces with 3, 4 and 5 cells, appeared in 2007 on a Web site run by Torsten Sillke who presumably found the construction. Checking Anneke on the Web, Knuth discovered she had invented several unusual mechanical puzzles which are marketed here by Kadon.

ANSWER

The only structure in Figure 29 that is impossible to construct with the seven Soma pieces is the skyscraper.

BIBLIOGRAPHY

Soma Puzzle Solutions. M. Wilson. Creative Publications, 1973.
"Soma Cubes." G. S. Carson in *Mathematics Teacher* (November 1973): 583–592.
Soma Cubes. S. Farhi. Pentacubes Puzzles Ltd., Auckland, Australia, 1979. This 15-page booklet depicts 111 Soma structures and their solutions.
"Soma: A Unique Object for Mathematical Study." D. Spector in *Mathematics Teacher* (May 1982): 404–407.
"The Computerized Soma Cube." J. Brunrell et al. in *Symmetry Unifying Human Understanding*, ed. I. Hargittai. Pergamon, 1987.
"Solving Soma Cube and Polyomino Problems Using a Microcomputer." D. A. Macdonald and Y. Gürsel in *Byte* (November 1989): 26–41.

CHAPTER SEVEN

Recreational Topology

TOPOLOGISTS HAVE BEEN called mathematicians who do not know the difference between a cup of coffee and a doughnut. Because an object shaped like a coffee cup can theoretically be changed into one shaped like a doughnut by a process of continuous deformation, the two objects are topologically equivalent, and topology can be roughly defined as the study of properties invariant under such deformation. A wide variety of mathematical recreations (including conjuring tricks, puzzles, and games) are closely tied to topological analysis. Topologists may consider them trivial, but for the rest of us they remain diverting.

A few years ago Stewart Judah, a Cincinnati magician, originated an unusual parlor trick in which a shoelace is wrapped securely around a pencil and a soda straw. When the ends of the shoelace are pulled, it appears to penetrate the pencil and cut the straw in half. The trick is disclosed here with Judah's permission.

Begin by pressing the soda straw flat and attaching one end of it, by means of a short rubber band, to the end of an unsharpened pencil (1 in Figure 38). Bend the straw down, and ask someone to hold the pencil with both hands so that the top of the pencil is tilted away from you at a 45-degree angle. Place the middle of the shoelace over the pencil (2); then cross the lace behind the pencil (3). Throughout the winding, whenever a crossing occurs, the same end – say end a – must always overlap the other end. Otherwise, the trick will not work.

Bring the ends forward, crossing them in front of the pencil (4). Bend the straw upward so that it lies along the pencil (5) and fasten its top end to the top of the pencil with another small rubber band. Cross the shoelace above the straw (6), remembering that

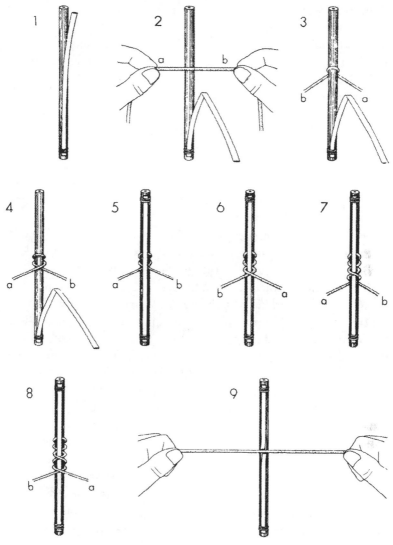

Figure 38. Stewart Judah's penetration trick. (Artist: Bunji Tagawa)

end *b* goes beneath end *a*. Wind the two ends behind the pencil for another crossing (7) and then forward for a final crossing in front (8). In these illustrations, the lace is spread out along the pencil to make the winding procedure clear. In practice, the windings may be tightly grouped near the middle of the pencil.

Ask the spectator to grip the pencil more firmly while you tighten the lace by tugging outward on its ends. Count three and give the ends a quick, vigorous pull. The last illustration in Figure 38 shows the surprising result. The shoelace pulls straight, apparently passing right through the pencil and slicing the straw, which (you explain) was too weak to withstand the mysterious penetration.

A careful analysis of the procedure reveals a simple explanation. Because the ends of the shoelace spiral around the pencil in a pair of mirror-image helices, the closed curve represented by performer and lace is not linked with the closed curve formed by spectator and pencil. The lace cuts the straw that holds the helices in place; then the helices annihilate each other as neatly as a particle of matter is annihilated by its antiparticle.

Many traditional puzzles are topological. In fact topology had its origin in Leonhard Euler's classic analysis in 1736 of the puzzle of finding a path over the seven bridges at Königsberg without crossing a bridge twice. Euler showed that the puzzle was mathematically identical with the problem of tracing a certain closed network in one continuous line without going over any part of the network twice. Route-tracing problems of this sort are common in puzzle books. Before tackling one of them, first note how many nodes (points that are the ends of line segments) have an even number of lines leading to them, and how many have an odd number. (There will always be an even number of "odd" nodes; cf., problem 8 in Chapter 5.) If all the nodes are "even," the network can be traced with a "re-entrant" path beginning anywhere and ending at the same spot. If two points are odd, the network can still be traced, but only if you start at one odd node and end at the other. If the puzzle can be solved at all, it can also be solved with a line that does not cross itself at any point. If there are more than two odd nodes, the puzzle has no solution. Such nodes clearly must be the end points of the line, and every continuous line has either two end points or none.

With these Eulerian rules in mind, puzzles of this type are easily solved. However, by adding additional features, such puzzles can often be transformed into first-class problems. Consider, for example, the network shown in Figure 39. All its nodes are even, so we know it can be traced in one re-entrant path. In this case, however, we permit any portion of the network to be retraced as often as desired, and you may begin at any point and end at any point. The

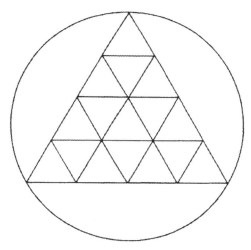

Figure 39. The network-tracing puzzle. (Artist: Bunji Tagawa)

problem: What is the minimum number of corner turns required to trace the network in one continuous line? Stopping and reversing direction is of course regarded as a turn.

Mechanical puzzles involving cords and rings often have close links with topological-knot theory. In my opinion, the best of such puzzles is the one pictured in Figure 40. It is easily made from a piece of heavy cardboard, string and any ring that is too large to pass through the central hole of the panel. The larger the cardboard and the heavier the cord, the easier it will be to manipulate the puzzle. The problem is simply to move the ring from loop A to loop B without cutting the cord or untying its ends.

This puzzle is described in many old puzzle books, usually in a decidedly inferior form. Instead of tying the ends of the cord to the panel, as shown here, each end passes through a hole and is fastened to a bead, which prevents the end from coming out of the hole. This permits an inelegant solution in which loop X is drawn through the two end holes and passed over the beads. The puzzle can be solved, however, by a neat method in which the ends of the cord play no role whatever. It is interesting to note that the puzzle has no solution if the cord is strung so that loop X passes over and under the double cord as shown in the illustration at upper right of Figure 40.

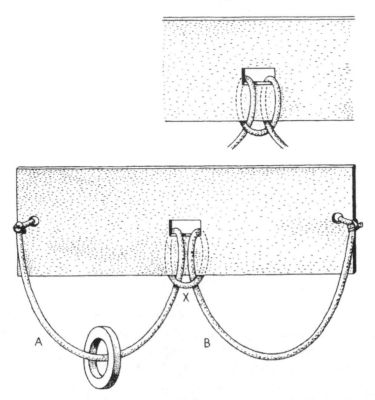

Figure 40. Can the ring be moved to loop B? (Artist: Bunji Tagawa)

Among the many mathematical games that have interesting topological features are the great Asian game of Go and the familiar children's game of "dots and boxes." The latter game is played on a rectangular array of dots, players alternately drawing a horizontal or vertical line to connect two adjacent dots. Whenever a line completes one or more unit squares, the player initials the square and plays again. After all the lines have been filled in, the player who has taken the most squares is the winner. The game can be quite exciting for skillful players because it abounds in opportunities for gambits in which squares are sacrificed in return for capturing a larger number later.

Although the game of dots and boxes is played almost as widely as ticktacktoe, no complete mathematical analysis of it has yet been published. In fact it is surprisingly complicated even on a square field as small as sixteen dots.

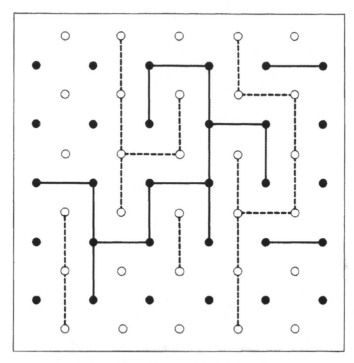

Figure 41. The topological game of Gale. (Artist: Bunji Tagawa)

David Gale, associate professor of mathematics at Brown University, has devised a delightful dot-connecting game, which I shall take the liberty of calling the game of Gale. It seems on the surface to be similar to the topological game of Hex explained in my Book 1. Actually it has a completely different structure (see Figure 41). The field is a rectangular array of black dots embedded in a similar rectangular array of colored dots. (In the illustration, colored dots are shown as circles and colored lines as dotted.) Player A uses a pencil with a black lead. On his turn he connects two adjacent black dots, either horizontally or vertically. His objective is a continuous line connecting the left and right sides of the field. Player B uses a colored pencil to join two adjacent colored dots. His objective is a line connecting the top and bottom of the field. No line is permitted to cross an opponent's line. Players draw one line only at each turn, and the winner is the first to complete a continuous line between his two sides of the field. The illustration depicts a winning game for the player with the colored pencil.

Gale can be played on fields of any size, though fields smaller than the one shown here are too easily analyzed to be of interest except to novices. It can be proved that the first player on any size board has the winning strategy; the proof is the same as the proof of first-player advantage in the game of Hex. Unfortunately, neither proof gives a clue to the nature of the winning strategy.

ADDENDUM

In 1960 the game of Gale, played on a board exactly like the one pictured here, was marketed by Hasenfield Brothers, Inc., Central Falls, RI, under the trade name of Bridg-it. Dots on the Bridg-it board are raised, and the game is played by placing small plastic bridges on the board to connect two dots. This permits an interesting variation, explained in the Bridg-it instruction sheet. Each player is limited to a certain number of bridges, say 10. If no one has won after all 20 bridges have been placed, the game continues by shifting a bridge to a new position on each move.

In 1951, seven years before Gale was described in my column, Claude E. Shannon (now professor of communications science and mathematics at the Massachusetts Institute of Technology) built the first Gale-playing robot. Shannon called the game Bird Cage. His machine plays an excellent, though not perfect, game by means of a simple computer circuit based on analog calculations through a resistor network. In 1958, another Gale-playing machine was designed by W. A. Davidson and V. C. Lafferty, two engineers then at the Armour Research Foundation of the Illinois Institute of Technology. They did not know of Shannon's machine, but based their plan on the same basic principle that Shannon had earlier discovered.

This principle operates as follows. A resistor network corresponds to the lines of play open to one of the players, say player A (see Figure 42). All resistors are of the same value. When A draws a line, the resistor corresponding to that line is short circuited. When B draws a line, the resistor, corresponding to A's line that is *intersected* by B's move, is open circuited. The entire network is thus

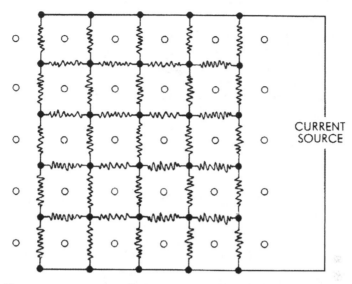

Figure 42. Resistor network for robot Gale player. (Artist: Bunji Tagawa)

shorted (i.e., resistance drops to zero) when A wins the game, and the current is cut off completely (i.e., resistance becomes infinite) when B wins. The machine's strategy consists of shorting or opening the resistor across which the maximum voltage occurs. If two or more resistors show the same maximum voltage, one is picked at random.

Actually, Shannon built two Bird Cage machines in 1951. In his first model the resistors were small light bulbs and the machine's move was determined by observing which bulb was brightest. Because it was often difficult to decide which of several bulbs was brightest, Shannon built a second model in which the bulbs were replaced by neon lamps and a network that permitted only one lamp to go on. When it goes on, a lockout circuit prevents any other lamp from lighting. Moves are made by switches that are all in intermediate positions at the start of the game. One player moves by closing a switch, the other by opening a switch.

When the machine has first move, Shannon reports, it almost always wins. Out of hundreds of games played, the machine has had only two losses when it had the first move, and they may have been due to circuit failures or improper playing of the game. If the human player has first move, it is not difficult to beat the machine, but the machine wins if a gross error is made.

ANSWERS

The figure-tracing puzzle can be solved with as few as 13 corner turns. Start at the second node from the left on the large triangle in Figure 39. Move up and to the right as far as possible, then left, then down and right to the base of the triangle, up and right, left as far as possible, down and right, right to the corner of the large triangle, up to the top of the triangle, down to the triangle's left corner, all the way around the circle, right to the third node on the triangle's base, up and left as far as possible, right as far as possible, then down and left to the base.

The cord-and-ring puzzle is solved as follows. Loosen the center loop enough so that the ring can be pushed up through it. Hold the ring against the front side of the panel while you seize the double cord where it emerges from the center hole. Pull the double cord toward you. This will drag a double loop out of the central hole. Pass the ring through this double loop. Now reach behind the panel and pull the double loop back through the hole so that the cord is restored to starting position. It only remains to slide the ring down through the center loop and the puzzle is solved.

POSTSCRIPT

Elwyn Berlekamp is the world's expert on dots and boxes. Although the game remains unsolved in general, much is known about the game when played on small boards. See Berlekamp's book *The Dots-and-Boxes Game: Sophisticated Child's Play* (A K Peters, 2000). He tells me that the game has been solved only through the 16-dot square, on which the second player can always win.

Figure 43 reproduces a picture from Sam Loyd's famous *Cyclopedia of Puzzles*. "What is the next best play," he asks, "and how many boxes will it win?" The game cannot end in a draw because there is an odd number of boxes. For Loyd's solution, see pages 152–153 of the *Mathematical Puzzles of Sam Loyd*, Vol. 1, which I edited (Dover, 1959).

THE BOXER'S PUZZLE

Figure 43. From Sam Loyd's *Cyclopedia of Puzzles.*

Soon after I introduced the game of Gale, a board version appeared on the market under the name of Bridg-it. The game was completely solved by Oliver Gross by a simple pairing strategy. The first player can always win. I give Gross's elegant solution in the chapter on Bridg-it and other games in Book 3. On Shannon's game and Bridg-it, and their many variants, see Cameron Browne's great book *Connection Games* (A K Peters, 2005).

BIBLIOGRAPHY

"Judah Pencil, Straw and Shoestring." Stewart Judah. An undated four-page typescript, issued by U. F. Grant, a magic dealer in Columbus, Ohio.

On the Tracing of Geometrical Figures. J. C. Wilson. Oxford University Press, 1905.

Puzzles Old and New. Professor Hoffmann (pseudonym of Angelo Lewis). Frederick Warne and Company, 1893.

The Dots-and-Boxes Game. Elwyn Berlecamp. A K Peters, 2000.

CHAPTER EIGHT

Phi: The Golden Ratio

PI, THE RATIO of the circumference of a circle to its diameter, is the best-known of all irrational numbers; that is, numbers with decimal expansions that are unending and nonrepeating. The irrational number phi (φ) is not so well-known, but it expresses a fundamental ratio that is almost as ubiquitous as pi, and it has the same pleasant propensity for popping up where least expected. (For example, see the discussion of the spot game in Chapter 13.)

A glance at the line in Figure 44 will make the geometrical meaning of phi clear. The line has been divided into what is commonly called the "golden ratio." The length of the line is to segment A as the length of segment A is to segment B. In each case the ratio is phi. If the length of B is 1, we can compute the value of phi easily from the following equation:

$$\frac{A+1}{A} = \frac{A}{1}$$

This can be written as the simple quadratic $A^2 - A - 1 = 0$, for which A has the positive value:

$$\frac{1 + \sqrt{5}}{2}$$

This is the length of A and the value of phi. Its decimal expansion is 1.61803398.... If the length of A is taken as 1, then B will be the reciprocal of phi ($1/\varphi$). Curiously, this value turns out to be 0.61803398.... Phi is the only positive number that becomes its own reciprocal by subtracting 1.

Like pi, phi can be expressed in many ways as the sum of an infinite series. The extreme simplicity of the following two examples underscores phi's fundamental character:

$$\varphi = 1 + \cfrac{1}{1 + \cfrac{1}{1 + \cfrac{1}{1 + \cfrac{1}{1 + \cdots}}}}$$

$$\varphi = \sqrt{1 + \sqrt{1 + \sqrt{1 + \sqrt{1 + \cdots}}}}$$

The ancient Greeks were familiar with the golden ratio; there is little doubt that it was consciously used by some Greek architects and sculptors, particularly in the structure of the Parthenon. The U.S. mathematician Mark Barr had this in mind 50 years ago when he gave the ratio the name of phi. It is the first Greek letter in the name of the great Phidias who is believed to have used the golden proportion frequently in his sculpture. Perhaps one reason why the Pythagorean brotherhood chose the pentagram or five-pointed star as the symbol of their order is the fact that every segment in this figure is in golden ratio to the next smaller segment.

Many medieval and Renaissance mathematicians, especially confirmed occultists such as Kepler, became intrigued by phi almost to the point of obsession. H. S. M. Coxeter, at the head of his splendid article on the golden ratio (see the bibliography for this chapter), quotes Kepler as follows: "Geometry has two great treasures: one is the theorem of Pythagoras; the other, the division of a line into extreme and mean ratio. The first we may compare to a measure of gold; the second we may name a precious jewel." Renaissance writers spoke of the ratio as a "divine proportion" or, following Euclid, as "extreme and mean ratio." The term "golden section" did not come into use until the nineteenth century.

A 1509 treatise by Luca Pacioli, entitled *De Divina Proportione* and illustrated by Leonardo da Vinci (a handsome edition was published in Milan in 1956), is a fascinating compendium of phi's appearances in various plane and solid figures. It is, for example,

Figure 44. The golden ratio: A is to B as A + B is to A. (Artist: James Egleson)

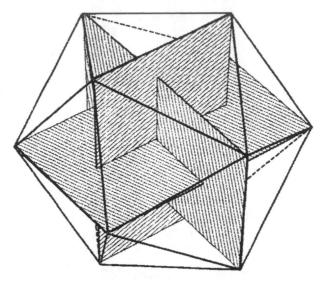

Figure 45. The corners of three golden rectangles coincide with the corners of an icosahedron. (Artist: James Egleson)

the ratio of the radius of a circle to the side of an inscribed regular decagon. And if we place three golden rectangles (rectangles with sides in golden ratio) so that they intersect each other symmetrically, each perpendicular to the other two, the corners of the rectangles will mark the 12 corners of a regular icosahedron as well as the centers of the 12 sides of a regular dodecahedron. (See Figures 45 and 46.)

The golden rectangle has many unusual properties. If we cut a square from one end, the remaining figure will be a smaller golden rectangle. We can keep snipping off squares, leaving smaller and smaller golden rectangles, as shown in Figure 47. (This is an example of a perfect squared rectangle of order infinity. See Chapter 17.) Successive points marking the division of sides into golden ratio lie on a logarithmic spiral that coils inward to infinity, its pole being the intersection of the two dotted diagonals. Of course these "whirling squares," as they have been called, can also be whirled outward to infinity by drawing larger and larger squares.

The logarithmic spiral is traceable in many other constructions involving phi. An elegant one makes use of an isosceles triangle that

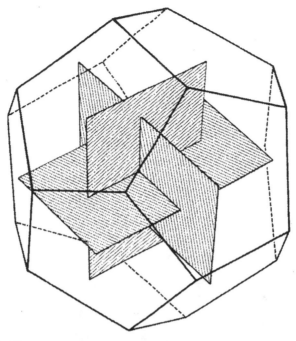

Figure 46. The corners of the same rectangles coincide with the centers of the sides of a dodecahedron. (Artist: James Egleson)

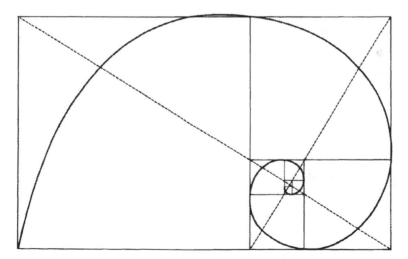

Figure 47. A logarithmic spiral indicated by "whirling squares." (Artist: James Egleson)

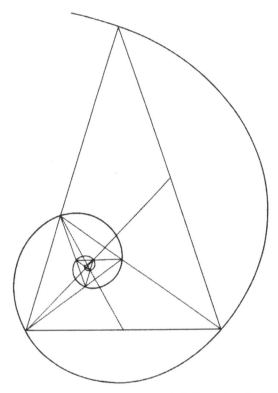

Figure 48. A logarithmic spiral indicated by "whirling triangles." (Artist: James Egleson)

has sides in golden ratio to its base (see Figure 48). Each base angle is 72 degrees, which is twice the top angle of 36 degrees. This is the golden triangle involved in the construction of the pentagram. If we bisect a base angle, the bisector cuts the opposite side in golden ratio to produce two smaller golden triangles, one of which is similar to the original. This triangle can in turn be divided by a base-angle bisector, and the process can be continued endlessly to generate a series of whirling triangles, which, like whirling squares, also stake out a logarithmic spiral. The pole of this spiral lies at the intersection of the two medians shown as dotted lines.

The logarithmic spiral is the only type of spiral that does not alter in shape as it grows, a fact that explains why it is so often found in nature. For example, as the mollusk inside a chambered nautilus grows in size, the shell enlarges along a logarithmic spiral so that it always remains an identical home. The center of a logarithmic spiral, viewed through a microscope, would look exactly like the spiral you would see if you continued the curve until it was as large as a galaxy and then viewed it from a vast distance.

The logarithmic spiral is intimately related to the Fibonacci series (1, 1, 2, 3, 5, 8, 13, 21, 34, . . .), in which every term is the sum of the two preceding terms. Biological growth often exhibits Fibonacci patterns. Commonly cited examples concern the spacing of leaves along a stalk and the arrangements of certain flower petals and seeds. Phi is involved here also, for the ratio between any two consecutive terms of the Fibonacci series comes closer and closer to phi as the series increases. Thus 5/3 is fairly close to phi (a 3 × 5 file card is hard to distinguish from a golden rectangle), but 8/5 is closer, and 21/13 is 1.619, which is closer still. In fact, if we start with any two numbers whatever and form an additive series (e.g., 7, 2, 9, 11, 20, . . .), the same convergence takes place. The higher the series goes, the closer the ratio between consecutive terms approaches phi.

This can be illustrated neatly by whirling squares. We begin with two small squares of any size, say the squares marked A and B in Figure 49. The side of square C is the sum of the sides of A and B. D is the sum of B and C, E is the sum of C and D, and so on. Regardless of the sizes of the two initial squares, the whirling squares get closer and closer to forming a golden rectangle.

There is a classic geometrical paradox that brings out strikingly how phi is linked to the Fibonacci series. If we dissect a square of 64 unit squares (see Figure 50), the four pieces can be put together again to make a rectangle of 65 square units. The paradox is explained by the fact that the pieces do not fit exactly along the long diagonal where there is a narrow space equal to one square unit. Note that the lengths of line segments in these figures are terms in a Fibonacci series. In fact, we can dissect the square so that these segments are consecutive terms in any additive series, and we will always get a form of the paradox, though in some cases the long rectangle will gain in area and in other cases it will lose area because

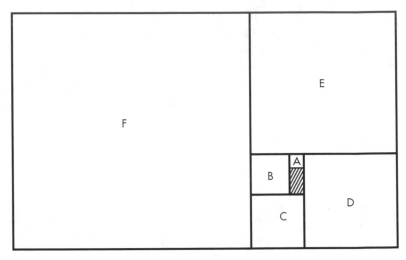

Figure 49. Squares show the convergence toward phi between consecutive terms in any additive series. (Artist: James Egleson)

of overlapping along the diagonal. This reflects the fact that consecutive terms in any additive series have a ratio that is alternately greater or less than phi.

The only way to cut the square so that there is no loss or gain of area in the rectangle is to cut it with segment lengths taken from the additive series $1, \varphi, \varphi + 1, 2\varphi + 1, 3\varphi + 2, \ldots$. Another way to write this series is $1, \varphi, \varphi^2, \varphi^3, \varphi^4, \ldots$. It is the only additive series in which the ratio between any two consecutive terms is constant. (The ratio

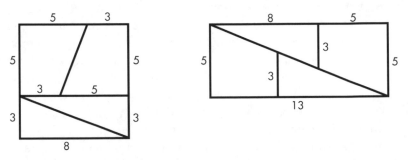

Figure 50. A paradox based on the properties of any additive series. (Artist: James Egleson)

is of course phi.) It is the golden series that all additive series strive vainly to become.

In recent times an enormous literature has developed around phi and related topics that is almost as eccentric as the circle-squaring literature revolving about pi. The classic is a 457-page German work, *Der goldene Schnitt*, written by Adolf Zeising and published in 1884. Zeising argues that the golden ratio is the most artistically pleasing of all proportions and the key to the understanding of all morphology (including human anatomy), art, architecture, and even music. Less crankish but comparable are Samuel Colman's *Nature's Harmonic Unity* (G. P. Putnam's Sons, 1912) and Sir Theodore Cook's *The Curves of Life* (Constable & Company, 1914).

Experimental esthetics may be said to have started with Gustav Fechner's attempts to give empirical support to Zeising's views. The great German psychologist measured thousands of windows, picture frames, playing cards, books, and other rectangles and checked the points at which graveyard crosses were divided. He found the average ratio close to phi. He also devised many ingenious tests in which subjects picked the most pleasing rectangle from a group, drew the most pleasing rectangle, placed the bar of a cross at the spot they liked best, and so on. Again, he found that preferences averaged close to phi. But his pioneer experiments were crude, and more recent work along similar lines has yielded only the cloudy conclusion that most people prefer a rectangle somewhere between a square and a rectangle that is twice as long as it is wide.

The American Jay Hambridge, who died in 1924, wrote many books defending what he called "dynamic symmetry," an application of geometry (with phi in a leading role) to art, architecture, furniture design, and even type fonts. Few today take his work seriously, though occasionally a prominent painter or architect will make deliberate use of the golden ratio in some way. George Bellows, for example, sometimes employed the golden ratio in planning the composition of a picture. Salvador Dali's *The Sacrament of the Last Supper* (owned by the National Gallery of Art, Washington, D. C.), is painted inside a golden rectangle, and other golden rectangles were used for positioning the figures. Part of an enormous dodecahedron floats above the table. (See Figure 51.)

Figure 51. *The Sacrament of the Last Supper.* Salvador Dali. National Gallery of Art, Washington, D. C. Chester Dale Collection.

Frank A. Lonc of New York has given considerable thought to phi. His booklets used to be obtainable from Tiffany Thayer's Fortean Society, which also peddled a German slide rule on which phi appears. (The Society did not continue after Thayer's death in 1959.) Lonc has confirmed one of Zeising's pet theories by measuring the heights of 65 women and comparing these figures to the heights of their navels, finding the ratio to average 1.618+. He calls this the Lonc Relativity Constant. "Subjects whose measurements did not fall within this ratio," he writes, "testified to hip-injuries or other deforming accidents in childhood." Lonc denies that the decimal expansion of pi is 3.14159..., as is widely believed. He has computed it more accurately by squaring phi, multiplying the result by 6, then dividing by 5 to get 3.14164078644620550.

I close with an interesting problem involving phi and the emblem made familiar by Charles de Gaulle, the two-beamed cross depicted in Figure 52. The cross is here formed of 13 unit squares. The problem is to draw a straight line through point A so that the total area on the shaded side of the line equals the area on the other side. Exactly how long is BC if the line is accurately placed? (In the illustration, the diagonal is incorrectly drawn so as to give no clue to its correct position.)

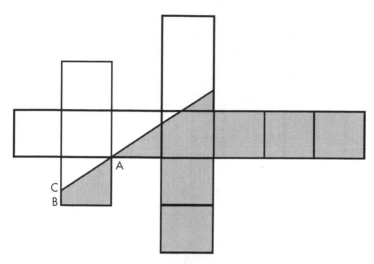

Figure 52. How long is the line BC? (Artist: James Egleson)

ADDENDUM

Many informative letters were received about the phi article. Several readers pointed out that in most mathematical books and journals the common symbol for the golden ratio is tau instead of phi. This is true, but phi is used in many crank books on the subject, and it is coming to be the symbol most often encountered in the literature of recreational mathematics. William Schaaf, for instance, uses it in his introductory remarks to the section on the golden ratio in his bibliographic work, *Recreational Mathematics*, published by the National Council of Teachers of Mathematics.

David Johnson, of the Philco Corporation, Palo Alto, California, used the firm's TRANSAC S-2000 computer to calculate phi to 2,878 decimal places. It took the machine a little less than four minutes to do the job. For numerologists I can report that the unusual sequence 177111777 occurs among the first 500 decimals.

L. E. Hough, a reader in Nome, Alaska, wrote to say that the two dotted diagonals in Figure 47, as well as the two dotted medians in Figure 48, are in golden ratio to each other.

Stephen Barr, whose father Mark Barr gave phi its name, sent me a clipping of an article by his father (in the London *Sketch*, about 1913) in which the concept of phi is generalized as follows. If we form a three-step series in which each term is the sum of the three previous terms, the terms approach a ratio of 1.8395+. A four-step series, each term the sum of four previous terms, approaches a ratio of 1.9275+. In general:

$$n = \frac{\log(2 - x)^{-1}}{\log x}$$

where n is the number of steps and x is the ratio that the series approaches. When n is 2, we have the familiar Fibonacci series in which x is phi. As n approaches infinity, x approaches 2.

Zeising's theory about navel heights continues to turn up in modern books. For example, in *The Geometry of Art and Life* by Matila Ghyka, published by Sheed and Ward in 1946, we read that "one can, in fact, state that if one measures the ratio for a great number of male and female bodies, the average ratio obtained will be 1.618." This makes about as much sense as computing the "average ratio" of the length of a bird's bill to the length of its leg. What group does one use for obtaining an average: people picked at random in New York, or Shanghai, or from the world population? To make things worse, the mixtures of body types in the world, or even in a small section of the world, is far from constant.

Kenneth Walters, of Seattle, and his friends took some measurements of the navel heights of their wives and arrived at an average ratio of 1.667, a bit higher than Lonc's 1.618. "Please understand," Walters wrote, "that our Hi-Phi wives were measured by their respective and respected husbands. It seems advisable that Mr. Lonc take up studies other than navel architecture."

Illinois, it has been noted, can call itself the "golden state" because its area code is 618, and its zip code starts with 618.

ANSWERS

The problem of bisecting the Gaullist cross can be solved algebraically by letting x be the length CD (see Figure 53) and y be the length MN. If the diagonal line bisects the cross, the shaded triangle must have an area of $2\frac{1}{2}$ square units. This permits us to write the equation $(x + 1)(y + 1) = 5$. Because triangles ACD and AMN are similar, we can also write the equation, $x/1 = 1/y$.

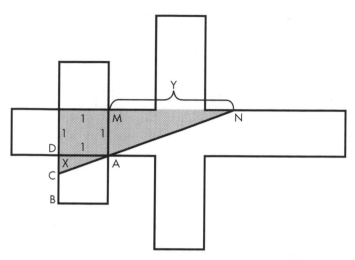

Figure 53. Solution to cross problem. (Artist: James Egleson)

The two equations combine to give x a value of $\frac{1}{2}(3 - \sqrt{5})$. BC therefore has a length of $\frac{1}{2}(\sqrt{5} - 1)$, or 0.618+, which is the reciprocal of phi $(1/\varphi)$. In other words, BD is divided by C in golden ratio. The lower end of the diagonal line similarly divides the side of the unit square in golden ratio. The bisecting line has a length of $\sqrt{15}$.

To find point C with compass and straightedge, we can adopt any of several simple methods that go back to Euclid. One is as follows.

Draw BE as shown in Figure 54. This bisects AD, making DF one half of BD. With the point of the compass at F, draw arc of circle with radius DF, intersecting BF at G. With point of compass at B, draw arc of circle with radius BG, intersecting BD at C. BD is now divided into the required golden ratio.

Several readers found easier ways to solve this problem. Nelson Max of Baltimore gave the simplest construction for the bisecting line. A semicircle, with one end at A (in Figure 53) and the other end at a point three units directly beneath A, intersects the right side of the cross at point N.

POSTSCRIPT

Although I have hinted that much of the literature on the golden ratio is crankish, I did not realize the extent to which this is true

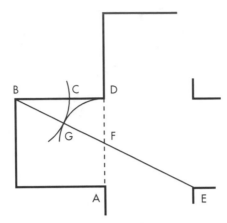

Figure 54. (Artist: James Egleson.)

until I read George Markowsky's 1992 article cited in the bibliography. It prompted me to write "The Cult of the Golden Ratio." This first appeared in the *Skeptical Inquirer* (Spring 1994), and was later reprinted in my *Weird Water and Fuzzy Logic* (Prometheus, 1996). For example, there is no reliable evidence that most people find the golden rectangle more pleasing than a file card's 3 × 5 ratio. See the references listed under "Attacks on Phi Addiction" in this chapter's bibliography.

Out of thousands of instances of phi turning up in geometric figures I select here three lovely examples. On the left of Figure 55 is a pentagram, symbol of the ancient Greek Pythagorean brotherhood and the pattern used by Goethe's Faust to trap Mephistopheles. Inverted, it is a traditional symbol of Satan. Every line segment in the star is in golden ratio to the next smaller segment.

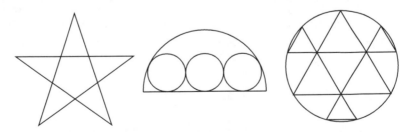

Figure 55. Golden ratio patterns. (Artist: Harold Jacobs.)

At the illustration's center the radius of the large circular arc divided by the diameter of a small circle is phi. And on the right, the side of a large triangle divided by the side of a small triangle is phi. Proofs of these assertions are pleasant exercises.

For more about the golden ratio see the chapter on Fibonacci numbers in Book 8.

BIBLIOGRAPHY

On Growth and Form. D'Arcy Wentworth Thompson. Cambridge University Press, 1917.

"The Golden Section, Phyllotaxis and Wythoff's Game." H. S. M. Coxeter in *Scripta Mathematica* 19 (June–September 1953): 135–143.

The Golden Number. Miloutine Borissavliévitch. Philosophical Library, 1958.

The Theory of Proportion in Architecture. P. H. Scholfield. Cambridge University Press, 1958.

"The Golden Section and Phyllotaxis." H. S. M. Coxeter in *Introduction to Geometry,* Chapter 11. John Wiley and Sons, 1961.

The Divine Proportion. H. E. Huntley, Dover, 1970.

The Golden Section and Related Curiosa. Garth E. Runion. Scott Foresman, 1972.

The Golden Ratio and Fibonacci Numbers. Richard A. Dunlap, Word Scientific, 1997.

A Mathematical History of the Golden Number. Roger Herz-Fischler. Dover, 1998.

The Golden Section. Hans Walser. The Mathematical Association of America, 2001.

The Golden Ratio. Mario Livio. Broadway Books, 2003.

The Golden Section: Nature's Greatest Secret. Scott Olsen. Walker, 2006.

ATTACKS ON PHI ADDITION

"Misconceptions About the Golden Ratio." George Markowsky in *The College Mathematics Journal* 21 (January 1993): 2–19.

"The Cult of the Golden Ratio." Martin Gardner in *Weird Water and Fuzzy Logic.* Prometheus, 1996.

"The Golden Ratio – A Contrary Viewpoint." Clement Falbo in *The College Mathematics Journal* 16 (November 2005): 123–134.

"The Golden Ratio." George Markowsky in *Notices of the American Mathematical Society* 52 (March 2005): 344–347. A critical review of the Livio book cited above.

"Bad News for Fibophiles." Miriam Abbott in *Philosophy Now* (February–March 2006): 32–33.

CHAPTER NINE

The Monkey and the Coconuts

IN THE OCTOBER 9, 1926, issue of *The Saturday Evening Post* appeared a short story by Ben Ames Williams entitled "Coconuts." The story concerned a building contractor who was anxious to prevent a competitor from getting an important contract. A shrewd employee of the contractor, knowing the competitor's passion for recreational mathematics, presented him with a problem so exasperating that while he was preoccupied with solving it, he forgot to enter his bid before the deadline.

Here is the problem exactly as the clerk in Williams's story phrased it:

> Five men and a monkey were shipwrecked on a desert island, and they spent the first day gathering coconuts for food. Piled them all up together and then went to sleep for the night.
>
> But when they were all asleep one man woke up, and he thought there might be a row about dividing the coconuts in the morning, so he decided to take his share. So he divided the coconuts into five piles. He had one coconut left over, and he gave that to the monkey, and he hid his pile and put the rest all back together.
>
> By and by the next man woke up and did the same thing. And he had one left over, and he gave it to the monkey. And all five of the men did the same thing, one after the other; each one taking a fifth of the coconuts in the pile when he woke up, and each one having one left over for the monkey. And in the morning they divided what coconuts were left, and they came out in five equal shares. Of course each one must have known there were coconuts missing; but each one was guilty as the others, so they didn't say anything. How many coconuts were there in the beginning?

Williams neglected to include the answer in his story. It is said that the offices of *The Saturday Evening Post* were showered with some 2,000 letters during the first week after the issue appeared. George Horace Lorimer, then editor-in-chief, sent Williams the following historic wire:

FOR THE LOVE OF MIKE, HOW MANY COCONUTS? HELL POPPING AROUND HERE.

For 20 years, Williams continued to receive letters requesting the answer or proposing new solutions. Today the problem of the coconuts is probably the most worked on and least often solved of all the Diophantine brain-teasers. (The term "Diophantine" is descended from Diophantus of Alexandria, a Greek algebraist who was the first to analyze extensively equations calling for solutions in rational numbers.)

Williams did not invent the coconut problem. He merely altered a much older problem to make it more confusing. The older version is the same except that in the morning, when the final division is made, there is again an extra coconut for the monkey; in Williams's version the final division comes out even. Some Diophantine equations have only one answer (e.g., $x^2 + 2 = y^3$); some have a finite number of answers; some (e.g., $x^3 + y^3 = z^3$) have no answer. Both Williams's version of the coconut problem and its predecessor have an infinite number of answers in whole numbers. Our task is to find the smallest positive number.

The older version can be expressed by the following six indeterminate equations, which represent the six successive divisions of the coconuts into fifths. N is the original number; F, the number each sailor received on the final division. The 1's on the right are the coconuts tossed to the monkey. Each letter stands for an unknown integer:

$$N = 5A + 1$$
$$4A = 5B + 1$$
$$4B = 5C + 1$$
$$4C = 5D + 1$$
$$4D = 5E + 1$$
$$4E = 5F + 1$$

It is not difficult to reduce these equations by familiar algebraic methods to the following single Diophantine equation with two unknowns:

$$1,024N = 15,625F + 11,529$$

This equation is much too difficult to solve by trial and error, and although there is a standard procedure for solving it by an ingenious use of continued fractions, the method is long and tedious. Here we shall be concerned only with an uncanny but beautifully simple solution involving the concept of *negative* coconuts. This solution is sometimes attributed to the University of Cambridge physicist P. A. M. Dirac (1902–1984), but in reply to my query Professor Dirac wrote that he obtained the solution from J. H. C. Whitehead, professor of mathematics (and nephew of the famous philosopher). Professor Whitehead, answering a similar query, said that he got it from someone else, and I have not pursued the matter further.

Whoever first thought of negative coconuts may have reasoned something like this. Since N is divided six times into five piles, it is clear that 5^6 (or 15,625) can be added to any answer to give the next highest answer. In fact any multiple of 5^6 can be added, and similarly any multiple can be subtracted. Subtracting multiples of 5^6 will of course eventually give us an infinite number of answers in negative numbers. These will satisfy the original equation, though not the original problem, which calls for a solution that is a positive integer.

Obviously there is no small positive value for N that meets the conditions, but possibly there is a simple answer in negative terms. It takes only a bit of trial and error to discover the astonishing fact that there is indeed such a solution: -4. Let us see how neatly this works out.

The first sailor approaches the pile of -4 coconuts, tosses a positive coconut to the monkey (it does not matter whether the monkey is given his coconut before or after the division into fifths), thus leaving five negative coconuts. These he divides into five piles, a negative coconut in each. After he has hidden one pile, four negative coconuts remain – exactly the same number that was there at the start! The other sailors go through the same ghostly ritual, the entire procedure ending with each sailor in possession of two negative coconuts, and the monkey, who fares best in this inverted operation, scurrying off happily with six positive coconuts. To find the answer that is the lowest positive integer, we now have only to add 15,625 to −4 to obtain 15,621, the solution we are seeking.

This approach to the problem provides us immediately with a general solution for n sailors, each of whom takes one nth of the coconuts at each division into $n$ths. If there are four sailors, we begin with three negative coconuts and add 4^5. If there are six sailors, we begin with five negative coconuts and add 6^7, and so on for other values of n. More formally, the original number of coconuts is equal to $k(n^{n+1}) - m(n - 1)$, where n is the number of men, m is the number of coconuts given to the monkey at each division, and k is an arbitrary integer called the parameter. When n is 5 and m is 1, we obtain the lowest positive solution by using a parameter of 1.

Unfortunately, this diverting procedure will not apply to Williams's modification, in which the monkey is deprived of a coconut on the last division. I leave it to the interested reader to work out the solution to the Williams version. It can of course be found by standard Diophantine techniques, but there is a quick shortcut if you take advantage of information gained from the version just explained. For those who find this too difficult, here is a very simple coconut problem free of all Diophantine difficulties.

Three sailors come upon a pile of coconuts. The first sailor takes half of them plus half a coconut. The second sailor takes half of what is left plus half a coconut. The third sailor also takes half of what remains plus half a coconut. Left over is exactly one coconut which they toss to the monkey. How many coconuts were there in the original pile? If you will arm yourself with 20 matches, you will have ample material for a trial-and-error solution.

ADDENDUM

If the use of negative coconuts for solving the earlier version of Ben Ames Williams's problem seems not quite legitimate, essentially the same trick can be carried out by painting four coconuts blue. Norman Anning, now retired from the mathematics department of the University of Michigan, hit on this colorful device as early as 1912 when he published a solution (*School Science and*

Mathematics, June 1912, page 520) to a problem about three men and a supply of apples. Anning's application of this device to the coconut problem is as follows.

We start with 5^6 coconuts. This is the smallest number that can be divided evenly into fifths, have one fifth removed and the process repeated six times, with no coconuts going to the monkey. Four of the 5^6 coconuts are now painted blue and placed aside. When the remaining supply of coconuts is divided into fifths, there will of course be one left over to give the monkey.

After the first sailor has taken his share, and the monkey has his coconut, we put the four blue coconuts back with the others to make a pile of 5^5 coconuts. This clearly can be evenly divided by 5. Before making this next division, however, we again put the four blue coconuts aside so that the division will leave an extra coconut for the monkey.

This procedure – borrowing the blue coconuts only long enough to see that an even division into fifths can be made and then putting them aside again – is repeated at each division. After the sixth and last division, the blue coconuts remain on the side, the property of no one. They play no essential role in the operation, serving only to make things clearer to us as we go along.

A good recent reference on Diophantine equations and how to solve them is *Diophantus and Diophantine Equations* by Isabella Bashmakova (The Mathematical Association of America, 1997).

There are all sorts of other ways to tackle the coconut problem. John M. Danskin, then at the Institute for Advanced Study, Princeton, NJ, as well as several other readers, sent ingenious methods of cracking the problem by using a number system based on 5. Scores of readers wrote to explain other unusual approaches, but all are a bit too involved to explain here.

ANSWERS

The number of coconuts in Ben Ames Williams's version of the problem is 3,121. We know from the analysis of the older version that $5^5 - 4$, or 3,121, is the smallest number that will permit five even divisions of the coconuts with one going to the monkey at each division. After these five divisions have been made, there will be

1,020 coconuts left. This number happens to be evenly divisible by 5, which permits the sixth division in which no coconut goes to the monkey.

In this version of the problem a more general solution takes the form of two Diophantine equations. When n, the number of men, is odd, the equation is

$$\text{Number of coconuts} = (1 + nk)n^n - (n - 1)$$

When n is even,

$$\text{Number of coconuts} = (n - 1 + nk)n^n - (n - 1)$$

In both equations k is the parameter that can be any integer. In Williams's problem, the number of men is 5, an odd number, so 5 is substituted for n in the first equation, and k is taken as 0 to obtain the lowest positive answer.

A letter from Dr. J. Walter Wilson, a Los Angeles dermatologist, reported an amusing coincidence involving this answer:

> Sirs:
>
> I read Ben Ames Williams's story about the coconut problem in 1926, spent a sleepless night working on the puzzle without success, then learned from a professor of mathematics how to use the Diophantine equation to obtain the smallest answer, 3,121.
>
> In 1939 I suddenly realized that the home on West 80th Street, Inglewood, California, in which my family and I had been living for several months, bore the street number 3121. Accordingly, we entertained all of our most erudite friends one evening by a circuit of games and puzzles, each arranged in a different room, and visited by groups of four in rotation.
>
> The coconut puzzle was presented on the front porch, with the table placed directly under the lighted house number blazingly giving the secret away, but no one caught on!

The simpler problem of the three sailors, at the end of the chapter, has the answer: 15 coconuts. If you tried to solve this by breaking matches in half to represent halves of coconuts, you may have concluded that the problem was unanswerable. Of course no coconuts need be split at all in order to perform the required operations.

Ben Ames Williams's story was reprinted in Clifton Fadiman's anthology, *The Mathematical Magpie* (1962), reissued in paperback by Copernicus in 1997. David Singmaster, in his unpublished history of famous mathematical puzzles, traces similar problems back to the Middle Ages. Versions appear in numerous puzzle books, as well as in textbooks that discuss Diophantine problems. My bibliography is limited to periodicals in English.

BIBLIOGRAPHY

"Solution to a Problem." Norman Anning in *School Science and Mathematics* (June 1912): 520.

"Solution to Problem 3,242." Robert E. Moritz in *The American Mathematical Monthly* 35 (January 1928): 47–48.

"The Problem of the Dishonest Men, the Monkeys, and the Coconuts." Joseph Bowden in *Special Topics in Theoretical Arithmetic*, pp. 203–212. Privately printed for the author by Lancaster Press, Inc., Lancaster, PA, 1936.

"Monkeys and Coconuts." Norman Anning in *The Mathematics Teacher* 54 (December 1951): 560–562.

"The Generalized Coconut Problem." R. B. Kirchner in *The American Mathematical Monthly* 67 (June–July 1960): 516–519.

"Five Sailors and a Monkey." P. W. Brashear in *The Mathematics Teacher* (October 1967): 597–599.

"On Coconuts and Integrity." T. Shin and G. Salvatore in *Crux Mathematicorum* 4 (August–September 1978): 182–185.

"On Dividing Coconuts: A Linear Diophantine Problem." S. Singh and D. Bhattacharya in *The College Mathematics Journal* (May 1997): 203–204.

"Coconuts – The History and Solution of a Classic Diophantine Problem." David Singmaster in *The Bulletin of the Indian Society for the History of Mathematics* 19 (1997): 35–51.

"More Coconuts." S. King in *The College Mathematics Journal* (September 1998): 312–313.

"Five Mathematicians, a Bunch of Coconuts, a Monkey, and a Coin." John E. Morrill in *The College Mathematics Journal* 35 (September 2004): 356–357.

CHAPTER TEN

Mazes

WHEN YOUNG THESEUS entered the Cretan labyrinth at Knossos in search of the dreaded Minotaur, he unwound a silken cord given him by Ariadne so that he could find his way out again. Architectural labyrinths of this sort – buildings with intricate passageways designed to bewilder the uninitiated – were not uncommon in the ancient world. Herodotus describes an Egyptian labyrinth that contained 3,000 chambers. Coins of Knossos bore a simple maze design, and more complicated maze patterns appeared on Roman pavements and on the robes of early Roman emperors. Throughout the Middle Ages the walls and floors of many cathedrals in Continental Europe were decorated with similar designs.

In England the most famous architectural labyrinth was Rosamond's Bower. It was reportedly built in a park at Woodstock in the twelfth century by King Henry II, who sought to conceal his mistress, Rosamond the Fair, from his wife, Eleanor of Aquitaine. Using Ariadne's string technique, goes the tale, Eleanor found her way to the center of the bower, where she forced the unhappy Rosamond to drink poison. The story caught the fancy of many writers – notably Joseph Addison, who wrote an opera about it, and Algernon Charles Swinburne, whose dramatic poem "Rosamond" is perhaps its most moving literary version.

Curiously, the Continental custom of decorating the interior of a cathedral with maze mosaics was not adopted in England. It was a common English practice, however, to cut mazes in the turf outside the church, where they were traversed as part of a religious ritual. These "quaint mazes in the wanton green," as Shakespeare called them, flourished in England until the eighteenth century. Garden mazes made of high hedges and intended solely for amusement

Figure 56. Plan of a hedge maze at Hampton Court. (Artist: Bunji Tagawa)

became fashionable during the late Renaissance. In England the most popular of the hedge mazes, through which confused tourists still wind their way, was designed in 1690 for the Hampton Court Palace of William of Orange. The present plan of the maze is reproduced in Figure 56.

The only hedge maze of historic significance in the United States was one constructed early in the nineteenth century by the Harmonists, a German Protestant sect which settled at Harmony, Indiana. (The town is now called New Harmony, the name given it in 1826 by the Scottish socialist Robert Owen, who established a Utopian colony there.) The Harmony labyrinth, like the medieval church mazes, symbolized the snakelike twists of sin and the difficulty of keeping on the true path. It was restored in 1941. Unfortunately no record of the original path had survived, so the restoration was made in an entirely new pattern.

From the mathematical standpoint a maze is a problem in topology. If its plan is drawn on a sheet of rubber, the correct path from entrance to goal is a topological invariant, which remains correct no matter how the rubber is deformed. You can solve a maze quickly on paper by shading all the blind alleys until only the direct routes remain. But when you are faced, as Queen Eleanor was, with the task of threading a maze of which you do not possess a map, it is a different matter. If the maze has one entrance, and the object is to find your way to the only exit, it can always be solved by placing your hand against the right (or left) wall and keeping it there as you walk. You are sure to reach the exit, though your route is not likely to be the shortest one. This procedure also works in the more traditional

Figure 57. A "simply connected" maze (left) and a "multiply connected" one (right). (Artist: Bunji Tagawa)

maze in which the goal is within the labyrinth, provided there is no route by which you can walk around the goal and back to where you started. If the goal is surrounded by one or more such closed circuits, the hand-on-wall method simply takes you around the largest circuit and back out of the maze; it can never lead you to the "island" inside the circuit.

Mazes that contain no closed circuits, such as the maze shown in the illustration at left in Figure 57, are called by topologists "simply connected." This is the same as saying that the maze has no detached walls. Mazes with detached walls are sure to contain closed circuits, and are known as "multiply connected" mazes (an example is depicted in the illustration at right). The hand-on-wall technique, used on simply connected mazes, will take you once in each direction along every path, so you are sure, somewhere along the route, to enter the goal. The Hampton Court maze is multiply connected, but its two closed loops do not surround the goal. The hand-on-wall technique will therefore carry you to the goal and back, but one corridor will be missed entirely.

Is there a mechanical procedure – an algorithm, to use a mathematical term – that will solve all mazes, including multiply connected ones with closed loops that surround the goal? There is, and the best formulation of it is given in Edouard Lucas's *Récréations mathématiques* (Volume I, 1882), where it is credited to M. Trémaux. As you walk through the maze, draw a line on one side of the path,

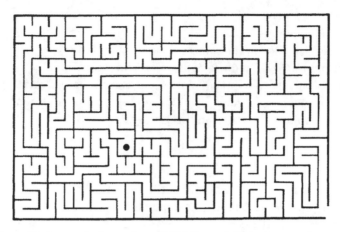

Figure 58. A maze in the garden of W. W. Rouse Ball. (Artist: Bunji Tagawa)

say your right. When you come to a new juncture of paths take any path you wish. If in walking along a new path you return to a previously visited juncture, or reach a dead end, turn around and go back the way you came. If in walking along an old path (a path marked on your left) you come to a previously visited juncture, take any new path, if one is available; otherwise, take an old path. Never enter a path marked on both sides.

The illustration at right in Figure 57 shows a multiply connected maze in which two closed circuits surround the central cell. If the reader will apply Trémaux's algorithm, using a red pencil to mark his trail, he will find that it will indeed take him to the center and back to the entrance after passing twice (once in each direction) through each portion of the maze. Better still, if you stop marking the paths once the goal is reached, you will have automatically recorded a direct route from entrance to goal. Simply follow the paths marked with one trail only.

For readers who might care to test this technique on a more difficult labyrinth, Figure 58 shows the plan of a multiply connected maze that the British mathematician W. W. Rouse Ball had traced out in his garden. The goal is the dot inside the maze.

Today's adults are no longer entertained by such puzzles, but there are two fields of science in which interest in mazes remains high: psychology and the designing of computers. Psychologists have of course been using mazes for several decades to study the learning behavior of men and animals. Even the lowly earthworm can be taught to run a maze of one fork, and the ant can learn mazes with as many as 10 points of choice. For computer designers, robot maze runners are part of an exciting program to build machines that, like animals, profit from their experience.

One of the earliest of these picturesque devices is Theseus, the famous maze-solving robot mouse invented by Claude E. Shannon, now at the Massachusetts Institute of Technology. (Theseus is an improvement on Shannon's earlier maze-solving "finger.") The "mouse" first works its way systematically through an unfamiliar maze, which may be multiply connected, by using a variation of Trémaux's algorithm. When the mouse reaches a juncture where it must make a choice, it does not do so in a random manner, as a man might, but it always takes the nearest path on a certain side. "It is rather difficult to trouble-shoot machines containing random elements," Shannon has explained. "It is difficult to tell when such a machine is misbehaving if you can't predict what it should do!"

Once the mouse has found its way to the goal, memory circuits enable it to run the maze a second time without error. In terms of Trémaux's system, this means that the mouse avoids all doubly traversed paths and tracks only the paths it has traveled once. This does not guarantee that it will take the shortest route to the goal, but only that it will reach the goal without entering any blind alleys. A real mouse is much slower in learning a maze because its exploration technique is largely (but not entirely) random trial and error, calling for many successes before the correct path is memorized.

Other robot maze runners have been built more recently. The most sophisticated, devised by Jaroslav A. Deutsch of the University of Oxford, is capable of transferring its training from one maze to another that is topologically equivalent even though its lengths and shapes have been altered. Deutsch's maze runner also takes advantage of short cuts added to the maze and does several other surprising things.

These devices are surely only crude beginnings. Future learning machines are likely to acquire enormous powers and to play unsuspected roles in the automatic machines of the space age. Mazes and space flight – the combination carries us back to the Greek myth mentioned at the beginning of this chapter. The maze of the Minotaur was built for King Minos by none other than Daedalus, who invented a pair of mechanical wings and whose son perished from flying too near the sun. "So cunningly contrived a mizmaze was never seen in the world, before nor since," writes Nathaniel Hawthorne in his *Tanglewood Tales* account of the story. "There can be nothing else so intricate, unless it were the brain of a man like Daedalus, who planned it, or the heart of any ordinary man."

POSTSCRIPT

Back in the 1970s, public interest in mazes reached an all-time high. Scores of books on the topic were published, ranging from simple mazes of the traditional type to a great variety of bizarre patterns. Robert Abbott's book *Mad Mazes* deserves special mention. Three-dimensional mazes appeared on the market in the form of transparent plastic cubes through which a marble is rolled from entrance to exit. An early example of such a maze, designed by Abbott, is described in Chapter 6 of my Book 5. *The Great Round the World Maze Trip* by Rick and Glory Brightfield (Ballantine, 1977) is based on a suggestion I proposed to Ballantine. The book – I wrote its introduction – contains mazes based on the streets of great cities.

If you enjoy solving traditional mazes, you can make the task more difficult by cutting a small hole in a sheet of paper, and then placing the paper on the maze so you see only the starting spot. You then try to find your way to the exit by moving the hole. In this way you play the role of persons struggling to find their way through such constructions as a hedge maze.

Chapter 6 in Book 5 considers algorithms for finding the shortest path through a maze and lists several papers on this task. One whimsical method requires modeling the maze with string.

Adrian Fisher, of England, is the world's top designer of both landscape mazes and paved floor mazes for zoos, churches, and other buildings. See his several beautiful books on mazes around the world, including those of his own construction. His article on "Paving Mazes" appears in *Puzzler's Tribute*, edited by David Wolfe and Tom Rodgers (A K Peters, 2002).

BIBLIOGRAPHY

HISTORY AND THEORY

"The Labyrinth of London." *The Strand Magazine* 35:208 (April 1908): 446. A reproduction of an old London map maze on which one attempts to enter by way of Waterloo Road and find his way to Saint Paul's Cathedral without crossing any road barriers.

The Labyrinth of New Harmony. Ross F. Lockridge. New Harmony Memorial Commission, 1941.

"Mazes and How to Thread Them." H. E. Dudeney in *Amusements in Mathematics*. Dover Publications, 1959.

"An Excursion into Labyrinths." Oystein Ore in *The Mathematics Teacher* (May 1959): 367–370.

Mazes and Labyrinths. W. H. Matthews. Dover, 1970.

Mazes and Labyrinths of the World. Janet Bond. Latimer, 1976.

Secrets of the Maze. Adrian Fisher and Howard Loxton. Barron's, 1997.

The Amazing Book of Mazes. Adrian Fisher. Abrams, 2006.

"Parity Puzzles." Adrian Fisher in *Games* (February 2007): 6–11. On the application of parity to a variety of unusual mazes.

MAZE-SOLVING COMPUTERS

"Presentation of a Maze-Solving Machine." Claude E. Shannon in *Cybernetics: Transactions of the Eighth Conference, March 15–16, 1951*: 173–180. Edited by Heinz von Foerster. Josiah Macy, Jr., Foundation, 1952.

"The Maze Solving Computer." Richard A. Wallace in *The Proceedings of the Association for Computing Machinery*, Pittsburgh (May 1952): 119–125.

"A Machine with Insight." J. A. Deutsch in *The Quarterly Journal of Experimental Psychology*, Vol. 6, Part I (February 1954): 6–11.

MAZE PUZZLES

For Amazement Only. Walter Shepherd. Penguin Books, no date; reissued by Dover Publications, and retitled *Mazes and Labyrinths*, in

1961. Fifty unusual mazes of all types. The author comments in detail on various psychological devices (including sexual symbols!) by which the astute maze maker can trick a solver into taking wrong paths. No discussion of mathematical theory, but a unique collection of difficult maze puzzles.

CHAPTER ELEVEN

Recreational Logic

> How often have I said to you that when you have eliminated the
> impossible, whatever remains, however improbable, must be the
> truth?
>
> Sherlock Holmes, *The Sign of Four*

A BRAIN-TEASER THAT calls for deductive reasoning with little or no
numerical calculation is usually labeled a logic problem. Of course
such problems are mathematical in the sense that logic may be
regarded as very general, basic mathematics; nevertheless, it is convenient to distinguish logic brain-teasers from their more numerous numerical cousins. Here we shall glance at three popular types
of recreational logic problems and discuss how to go about tackling
them.

The most frequently encountered type is sometimes called by
puzzlists a "Smith-Jones-Robinson" problem after an early brain-
teaser devised by the English puzzle expert Henry Dudeney (see his
Puzzles and Curious Problems, Problem 49). It consists of a series
of premises, usually about individuals, from which one is asked to
make certain deductions. A recent American version of Dudeney's
problem goes like this:

1. Smith, Jones, and Robinson are the engineer, brakeman, and
   fireman on a train, but not necessarily in that order. Riding
   the train are three passengers with the same three surnames,
   to be identified in the following premises by a "Mr." before
   their names.
2. Mr. Robinson lives in Los Angeles.
3. The brakeman lives in Omaha.

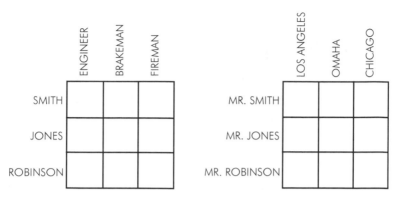

Figure 59. Two matrices for the Smith-Jones-Robinson problem. (Artist: Amy Kasai)

4. Mr. Jones long ago forgot all the algebra he learned in high school.
5. The passenger whose name is the same as the brakeman's lives in Chicago.
6. The brakeman and one of the passengers, a distinguished mathematical physicist, attend the same church.
7. Smith beat the fireman at billiards.

Who is the engineer?

It is possible to translate this problem into the notation of symbolic logic and solve it by appropriate techniques, but this is needlessly cumbersome. On the other hand, it is difficult to grasp the problem's logical structure without some sort of notational aid. The most convenient device to use is a matrix with vacant cells for all possible pairings of the elements in each set. In this case there are two sets and therefore we need two such matrices (see Figure 59).

Each cell is to be marked with a "1" to indicate that the combination is valid, or "0" to indicate that it is ruled out by the premises. Let us see how this works out. Premise 7 obviously eliminates the possibility that Smith is the fireman, so we place a "0" in the upper right corner cell of the matrix at left. Premise 2 tells us that Mr. Robinson lives in Los Angeles so we place a "1" in the lower left corner of the matrix on the right, and "0's" in the other cells of the same row and the same column to show that Mr. Robinson doesn't live in Omaha

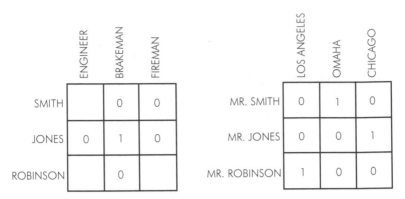

Figure 60. The matrices in use. (Artist: Amy Kasai)

or Chicago and that Mr. Smith and Mr. Jones do not live in Los Angeles.

Now we have to do a bit of thinking. Premises 3 and 6 inform us that the physicist lives in Omaha, but what is his name? He cannot be Mr. Robinson, nor can he be Mr. Jones (who has forgotten his algebra), so he must be Mr. Smith. We indicate this with a "1" in the middle cell of the top row in the matrix at right, and "0's" in the remaining empty cells of the same row and column. Only one cell in the matrix is now available for the third "1," proving that Mr. Jones lives in Chicago. Premise 5 now permits us to identify the brakeman as Jones, so we place a "1" in the central cell of the left-hand matrix and "0's" in the other cells of the same row and column. The appearance of our matrices at this stage is shown in Figure 60.

The remaining deductions are obvious. Only the bottom cell of the fireman's column is available for a "1." This puts a "0" in the lower left corner, leaving vacant only the top left corner cell for the final "1" which proves that Smith is the engineer.

Lewis Carroll was fond of inventing quaint and enormously complicated problems of this sort. Eight are to be found in the appendix of his *Symbolic Logic*. One monstrous Carrollian problem (involving 13 variables and 12 premises from which one is to deduce that no magistrates are snuff-takers) was fed to an IBM 704 computer by John G. Kemeny, chairman of the mathematics department at Dartmouth College. The machine solved the problem in about four minutes, although a complete printing of the problem's "truth table" (a matrix showing the validity or invalidity of every possible combination of true and false values for the variables) would have taken 13 hours!

For readers who care to try their luck on a more difficult Smith-Jones-Robinson problem, here is a new one devised by Raymond Smullyan of the mathematics department at Princeton University.

1. In 1918, on the day that the armistice of World War I was signed, three married couples celebrated by having dinner together.
2. Each husband is the brother of one of the wives, and each wife is the sister of one of the husbands; that is, there are three brother–sister pairs in the group.
3. Helen is exactly 26 weeks older than her husband, who was born in August.
4. Mr. White's sister is married to Helen's brother's brother-in-law. She (Mr. White's sister) married him on her birthday, which is in January.
5. Marguerite White is not as tall as William Black.
6. Arthur's sister is prettier than Beatrice.
7. John is 50 years old.
8. What is Mrs. Brown's first name?

Another familiar type of logic poser may be called the "colored-hat" variety after the following best-known example. Three men – A, B, and C – are blindfolded and told that either a red or a green hat will be placed on each of them. After this is done, the blindfolds are removed; the men are asked to raise a hand if they see a red hat, and to leave the room as soon as they are sure of the color of their own hat. All three hats happen to be red, so all three men raise a hand. Several minutes go by until C, who is more astute than the others, leaves the room. How did he deduce the color of his hat?

C asks himself: Can my hat be green? If so, then A will know immediately that he has a red hat for only a red hat on his head would cause B to lift his hand. A would therefore leave the room. B would reason the same way and also leave. Since neither has left, C deduces that his own hat must be red.

As George Gamow and Marvin Stern point out in their delightful little book *Puzzle-Math*, this can be generalized to any number of men who are all given red hats. Suppose there is a fourth man, D, who is more astute than C. He reasons that if his hat is green, then A, B, and C are in a situation exactly like the one just described. After several minutes the most astute member of the trio will surely leave the room. But if five minutes go by and no one leaves, D can deduce that his hat is red. If there is a fifth man more astute than D, he will decide that his hat is red after a time lapse of, say, 10 minutes. Of course all this is weakened by the assumption of different levels of astuteness and by vagueness about the length of the various time lapses.

Less ambiguous are some other colored-hat problems such as the following, also invented by Smullyan. Three men – A, B, and C – are aware that all three of them are "perfect logicians" who can instantly deduce all the consequences of a given set of premises. There are four red and four green stamps available. The men are blindfolded and two stamps are pasted on each man's forehead. The blindfolds are removed. A, B, and C are asked in turn: "Do you know the colors of your stamps?" Each says: "No." The question is then asked of A once more. He again says: "No." B is now asked the question, and replies: "Yes." What are the colors of B's stamps?

A third class of popular logic puzzles involves truth-telling and lying. The classic example concerns an explorer in a region inhabited by the usual two tribes; the members of one tribe always lie, the members of the other always tell the truth. He meets two natives. "Are you a truth-teller?" he asks the tall one. "Goom," the native replies. "He say 'Yes,'" explains the short native, who speaks English, "but him big liar." What tribe did each belong to?

A systematic approach would be to jot down the four possibilities – TT, TL, LT, LL – then eliminate the pairs that are inconsistent with the premises. A quicker solution is reached if one has the insight to see that the tall native must answer "Yes" regardless of whether he lies or tells the truth. Since the short native told the truth, he must be a truth-teller and his companion a liar.

The most notorious problem of this type, complicated by probability factors and semantic obscurity, was dropped casually by the British astronomer Sir Arthur Eddington into the middle of the sixth chapter of his *New Pathways in Science*. "If A, B, C, D each speak the truth once in three times (independently), and A affirms that B denies that C declares that D is a liar, what is the probability that D was speaking the truth?"

Eddington's answer of 25/71 was greeted by howls of protest from his readers, touching off a droll and confusing controversy that was never decisively resolved. The English astronomer Herbert Dingle, reviewing Eddington's book in *Nature* (March 23, 1935), dismissed the problem as meaningless and symptomatic of Eddington's confused thinking about probability. Theodore Sterne, an American physicist, replied (*Nature*, June 29, 1935) that the problem was not meaningless, but lacked sufficient data for a solution.

Dingle responded (*Nature*, September 14, 1935) by contending that, if one granted Sterne's approach, there were enough data to reach a solution of exactly 1/3. Eddington then reentered the fray with a paper entitled "The Problem of A, B, C, and D" (*The Mathematical Gazette*, October 1935), in which he explained in detail how he had calculated his answer. The controversy terminated with two articles in the same magazine (*The Mathematical Gazette*, December 1936), one defending Eddington and the other taking a position differing from all former ones.

The difficulty lies chiefly in deciding exactly how to interpret Eddington's statement of the problem. If B is truthful in making his denial, are we justified in assuming that C said that D spoke the truth? Eddington thought not. Similarly, if A is lying, can we then be sure that B and C said anything at all? Fortunately we can sidestep all these verbal difficulties by making (as Eddington did not) the following assumptions:

1. All four men made statements.
2. A, B, and C each made a statement that either affirmed or denied the statement that follows.
3. A lying affirmation is taken to be a denial and a lying denial is taken to be an affirmation.

The men lie at random, each averaging two lies out of every three statements. If we represent each man's true statement by T and his two lies by L1 and L2, we can construct a table of 81 different combinations of T's and L's for the four men. We must then decide which of these combinations are made impossible by the logical structure of the statement. The number of possible combinations terminating in T (i.e., ending with a true statement by D) divided by the total number of possible combinations will then be our answer.

ADDENDUM

In giving the problem about the explorer and the two natives, I should have made it more precise by saying that the explorer recognized the word "Goom" as a native word meaning either yes or no, but that he didn't know which. This would have forestalled a number of letters, such as the following one from John A. Jonelis of Indianapolis:

> Sirs:
>   I enjoyed the article on logic brain teasers. . . . Wishing to share this enjoyment with my wife, and probably to indulge my male ego, I teased her with the truth-teller–liar puzzle. Within two minutes she had a completely sound answer, diametrically opposed to your published one.
>   The tall native apparently cannot understand any English or he would be able to answer yes or no in English. His "Goom," therefore, meant something like "I do not understand" or "Welcome to Bongo Bongo land." Consequently, the small native was lying when he said his companion answered yes, and being a liar, lied when he called his companion a liar. The tall native is therefore a truth-teller.
>   This female logic threw my male ego for a loop. Does it deflate yours a bit?

ANSWERS

The first logic problem is best handled by three matrices: one for combinations of first and last names of wives, one for first and last names of husbands, and one to show sibling relationships. Since Mrs. White's first name is Marguerite (premise 5), we have only two alternatives for the names of the other wives: (1) Helen Black and Beatrice Brown or (2) Helen Brown and Beatrice Black.

Let us assume the second alternative. White's sister must be either Helen or Beatrice. It cannot be Beatrice, because then Helen's brother would be Black; Black's two brothers-in-law would be White (his wife's brother) and Brown (his sister's husband); but Beatrice Black is not married to either of them, a fact inconsistent with premise 4. Therefore, White's sister must be Helen. This in turn allows us to deduce that Brown's sister is Beatrice and Black's sister is Marguerite.

Premise 6 leads to the conclusion that Mr. White's first name is Arthur (Arthur Brown is ruled out because that would make Beatrice prettier than herself, and Arthur Black is ruled out because we know from premise 5 that Black's first name is William). Therefore Brown's first name must be John. Unfortunately premise 7 informs us that John was born in 1868 (50 years before the Armistice), which is a leap year. This would make Helen older than her husband by one day more than the 26 weeks specified in premise 3. (Premise 4 tells us that her birthday is in January, and premise 3 tells us her husband's birthday is in August. She can be exactly 26 weeks older than he only if her birthday is January 31, his on August 1, and there is no February 29 in between!) This eliminates the second of the two alternatives with which we started, forcing us to conclude that the wives are Marguerite White, Helen Black and Beatrice Brown. There are no inconsistencies because we do not know the year of Black's birth. The premises permit us to deduce that Marguerite is Brown's sister, Beatrice is Black's sister, and Helen is White's sister, but leave undecided the first names of White and Brown.

In the problem of the stamps on the foreheads, B has three alternatives: his stamps are (1) red-red, (2) green-green, or (3) red-green. Assume they are red-red.

After all three men have answered once, A can reason as follows: "I cannot have red-red (because then C would see four red stamps and know immediately that he had green-green, and if C had green-green, B would see four green stamps and know that he had red-red). Therefore I must have red-green."

But when A was asked a second time, he did not know the color of his stamps. This enables B to rule out the possibility that his own stamps are red-red. Exactly the same argument enables B to eliminate the possibility that his stamps are green-green. This leaves for him only the third alternative: red-green.

A dozen readers were quick to point out that there is a quick way to solve this problem without bothering to analyze any of the questions and answers! Brockway McMillan of Summit, New Jersey, expressed it this way:

> The statement of the problem is completely symmetrical as regards red and green stamps. Therefore, any distribution of stamps on foreheads which satisfies the stated conditions will, if red and green are interchanged, again become a distribution satisfying the conditions. Therefore, if the solution is unique, it must be invariant under the interchange of red and green. The only such solution is that B have a red and a green stamp.

As Wallace Manheimer, chairman of the mathematics department of a high school in Brooklyn, put it, this short-cut approach is based not on the fact that A, B, and C are perfect logicians, as stated in the problem, but on the fact that Raymond Smullyan is!

The answer to Eddington's problem of the four men is 13/41 as the probability that D is telling the truth. All combinations of truth-telling and lying that have an odd number of lies (or truths) prove to be inconsistent with Eddington's statement. This eliminates from the table of 81 possible combinations all but 41, of which 13 end with a true statement by D. Because each of the other three men is telling the truth in exactly the same number of valid combinations, the probability of having told the truth is the same for all four men.

Using the symbol of equivalence ($\equiv$), which means that the statements connected by the symbol are either both true or both false, and the symbol of negation ($\sim$), we can write Eddington's problem in the propositional calculus of symbolic logic as follows:

$$A \equiv [B \equiv \sim(C \equiv \sim D)]$$

This can be simplified to

$$A \equiv [B \equiv (C \equiv D)]$$

The truth table for this expression will confirm the results given in the previous analysis.

POSTSCRIPT

Since this chapter ran in *Scientific American*, Raymond Smullyan has become the world's most prolific composer of remarkable logic puzzles, especially of the liar and truth-teller variety. See Chapter 20, "Raymond Smullyan on Logic Puzzles," in my Book 13 and the references cited.

Smullyan is noted for his technical writings on formal logic and set theory, and for many simplifications of Gödel's famous proofs. In addition to his books on logic problems, Smullyan has also published collections of ingenious chess problems unlike any such problems previously devised.

Smullyan's subtle sense of humor pervades all his work, especially his philosophical essays. He likes to tell of the time a waitress asked Descartes if he wanted a cocktail. "I think not," he replied.

I love Smullyan's parable about a dream he once had in which all of the world's greatest philosophers appeared before him and gave precise compressed accounts of their philosophical systems. In the dream, Smullyan made a single remark that totally demolished each system. One by one, the philosophers, starting with Plato and Aristotle, left the scene in great embarrassment. Fearing he would never recall what he said, Smullyan jotted down his remark, then went back to sleep.

The next morning Smullyan could not remember what he had said, but he found what he had written. The remark was "That's what *you* say!"

For complex variations on the problem of the three men and the colored hats, see Chapter 10 of my Book 13 and the references cited. On Lewis Carroll, see Chapter 4 of my Book 3 and my book *The Universe in a Handkerchief* (Copernicus, 1996). It covers all of Carroll's mathematical recreations including his whimsical logic puzzles.

My book *Logic Machines and Diagrams* (there is a Dover paperback reprint) covers the use of diagrams, such as the Venn circles, for solving syllogisms and problems in the propositional calculus. It also includes Carroll's method of solving syllogisms by putting counters on a diagram as explained in his book *The Game of Logic*. Although Carroll did not realize it, his method applies even more efficiently to the propositional calculus. My book also gives a neat way to solve propositional calculus problems with directed graphs. On this I collaborated with graph theorist Frank Harary. Our paper earned me (for the second time) an Erdös number of 2.

BIBLIOGRAPHY

"Eddington's Probability Paradox." H. Wallis Chapman in *The Mathematical Gazette* 20: 241 (December 1936): 298–308.

Question Time. Hubert Phillips. Farrar and Rinehart, 1938.

An Experiment in Symbolic Logic on the IBM 704. John G. Kemeny. Rand Corporation Report P-966, September 7, 1956. Kemeny explains how the computer was programmed to solve the Lewis Carroll problem. This problem does not appear in Carroll's published writings but may be found in John Cook Wilson's *Statement and Inference*, Vol. 2, page 638, Oxford University Press, 1926. Wilson does not give the answer. The problem was first solved by L. J. Russell, using short-cut techniques of symbolic logic. See his article, "A Problem of Lewis Carroll," *Mind* 60: 239 (July 1951): 394–396.

101 Puzzles in Thought and Logic. Clarence Raymond Wylie, Jr. Dover Publications, 1957.

CHAPTER TWELVE

Magic Squares

THE TRADITIONAL MAGIC SQUARE is a set of integers in serial order, beginning with 1, arranged in square formation so that the total of each row, column and main diagonal is the same. Some notion of the fantastic lengths to which this largely frivolous topic has been analyzed may be gained from the fact that in 1838, when much less was understood about magic squares than is known today, a French work on the subject ran to three volumes. From ancient times until now the study of magic squares has flourished as a kind of cult, often with occult trappings, whose initiates range from such eminent mathematicians as Arthur Cayley and Oswald Veblen to laymen such as Benjamin Franklin.

The "order" of a magic square is the number of cells on one of its sides. There are no magic squares of order two, and only one (not counting its rotations and reflections) of order three. An easy way to remember this square is as follows: First write the digits in order as shown on the left in Figure 61; then move each corner digit to the far side of the central digit as indicated by the arrows. The result is the magic square shown on the right, which has a constant of 15. (The constant is always half the sum of n^3 and n, where n is the order.) In China, where this square is called the *lo-shu*, it has a long history as a charm. Today it is still found on amulets worn in the Far East and India, and on many large passenger ships it is the pattern for games of shuffleboard.

Magic squares grow quickly in complexity when we turn to order four. There are exactly 880 different types, again ignoring rotations and mirror images, many of which are much more magical than required by the definition of a magic square. One interesting

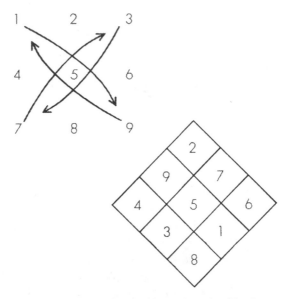

Figure 61. How the *lo-shu* can be formed. (Artist: Alex Semenoick)

species, known as a symmetrical square, appears in Albrecht Dürer's famous engraving *Melencolia* (see Figure 62).

Dürer never explained the rich symbolism of this masterpiece, but most authorities agree that it depicts the sullen mood of the thinker unable to engage in action. Today we call such a mood a "clinical depression." In the Renaissance, the melancholy temperament was thought characteristic of creative genius; it was the affliction of scholars "sicklied o'er with the pale cast of thought." (This notion that brilliant intellects are unable, like Hamlet, to make decisions is still with us; witness Harry Truman's public criticism of Adlai Stevenson on precisely such grounds.)

In Dürer's picture, unused tools of science and carpentry lie in disorder about the disheveled, brooding figure of Melancholy. There is nothing in the balance scales, no one mounts the ladder, the sleeping hound is half-starved, the winged cherub waits for dictation while time is running out in the hourglass above. The wooden sphere and curiously truncated stone tetrahedron suggest the mathematical base of the building arts. Apparently the scene is bathed in moonlight. The lunar rainbow arching over what appears to be a comet may signify the hope that the somber mood will pass.

Figure 62. Albrecht Dürer's *Melencolia*. At upper right is a magic square. (Image courtesy of Owen Gingerich)

Giorgio de Santillana, in his book *The Age of Adventure*, sees in this strange picture "the mysterious wondering pause of the Renaissance mind at the threshold of the as-yet-only-dreamt-of powerhouse of Science." James Thomson concludes his great poem of pessimism, *The City of Dreadful Night*, with a magnificent

twelve-stanza description of this picture, seeing in it a "confirmation of the old despair."

> The sense that every struggle brings defeat
> Because Fate holds no prize to crown success;
> That all the oracles are dumb or cheat
> Because they have no secret to express;
> That none can pierce the vast black veil uncertain
> Because there is no light beyond the curtain;
> That all is vanity and nothingness.

Fourth-order magic squares were linked to Jupiter by Renaissance astrologers and were believed to combat melancholy (which
was Saturnian in origin). This may explain the square in the upper
right-hand corner of the engraving. It is called symmetric because
each number added to the number symmetrically opposite the
square's center yields 17. Owing to this fact, there are many four-
cell groups (in additions to rows, columns, and main diagonals)
that total the fourth-order constant of 34; for example, the four corner cells, the four central cells, the 2 × 2 squares at each corner.
A square of this type can be constructed by an absurdly simple
method. Merely write in square array and in serial order the numbers 1 to 16, and then invert the two main diagonals. The result is
a symmetrical magic square. Dürer interchanged the two middle
columns of this square (which does not affect its properties) so that
the two middle cells of the bottom row would indicate the year he
made the engraving.

A fourth-order square, found in an eleventh- or twelveth-century
inscription at Khajuraho, India, is shown at the top of Figure 63. It
belongs to a species know as diabolic squares (also called "pandiagonal" and "Nasik"), which are even more astonishing than the symmetrical ones. In addition to the usual properties, diabolic squares
are also magic along all "broken diagonals." For example, cells 2,
12, 15, and 5, and cells 2, 3, 15, and 14, are broken diagonals that
can be restored by putting two duplicate squares alongside each
other. A diabolic square remains diabolic if a row is shifted from
top to bottom or bottom to top, and if a column is moved from
one side to the other. If we form a mosaic by fitting together a large

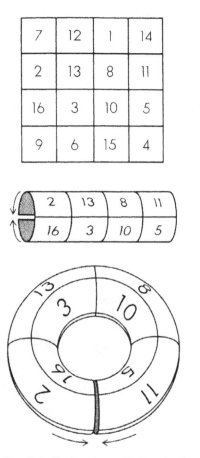

7	12	1	14
2	13	8	11
16	3	10	5
9	6	15	4

Figure 63. The diabolic doughnut. (Artist: Alex Semenoick)

number of duplicate diabolic squares, we have a field on which any 4 × 4 group of cells will be diabolic. Any four adjacent cells on the field, up and down, left and right, or diagonally, will yield the constant.

Perhaps the most dramatic way of exhibiting the diabolic properties of such a square is described by mathematicians J. Barkley Rosser and Robert J. Walker, both of Cornell University, in a paper published in 1938. We simply bring together the top and bottom of the square to make a cylinder, and then stretch and bend the cylinder into a torus (see Figure 63). All rows, columns, and diagonals

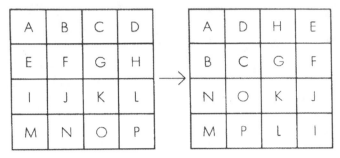

Figure 64. One of five transformations which do not destroy the diabolism of a diabolic square. (Artist: Alex Semenoick)

now become closed loops. If we start at any cell and move two squares away in any direction along a diagonal, we always arrive at the same cell. This cell is called the "antipode" of the cell where we began. Every pair of antipodes on this diabolic doughnut will total 17. Every loop of four cells, diagonally or orthogonally, adds up to 34, as does any square group of four cells.

A diabolic square remains diabolic under five different transformations: (1) a rotation, (2) a reflection, (3) a transfer of a row from top to bottom or vice versa, (4) a transfer of a column from one side to the other, and (5) a rearrangement of cells according to the plan shown in Figure 64. By combining these five transformations one can obtain 48 basic types of diabolic squares (384 if rotations and reflections are included). Rosser and Walker show that these five transformations constitute a "group" (an abstract structure with certain properties) that is identical with the group of transformations of the hypercube (four-dimensional cube) into itself.

The relation of diabolic squares to the hypercube is easily seen by transferring the 16 cells of such a square to the 16 corners of a hypercube. This can be shown on the familiar two-dimensional projection of a hypercube (see Figure 65). The sum of the four corners of each of the 24 square faces of this hypercube will be 34. The antipodal pairs, which add up to 17, are the diagonally opposite corners of the hypercube. By rotating and reflecting the hypercube, it can be placed in exactly 384 different positions, each of which maps back to the plane as one of the 384 diabolic squares.

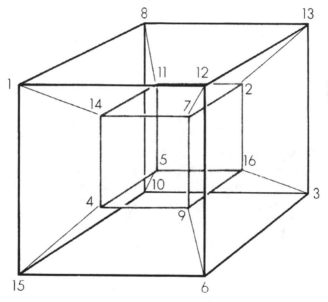

1	8	13	12
14	11	2	7
4	5	16	9
15	10	3	6

Figure 65. Diabolic hypercube and one of its 384 diabolic squares. (Artist: Alex Semenoick)

Claude Fayette Bragdon, a prominent U.S. architect and occultist who died in 1946, was fascinated by his discovery that on most magic squares a line traced from cell to cell in serial order will produce an artistically pleasing pattern. Other patterns can be found by tracing only the odd or only the even cells. Bragdon used "magic lines" obtained in this manner as a basis for textile patterns, book

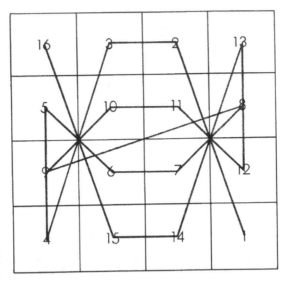

Figure 66. The "magic line" of Dürer's square. (Artist: Alex Semenoick.)

covers, architectural ornaments, and the decorative chapter head-
ings of his autobiography *More Lives Than One*. His design for
the ventilating grill in the ceiling of the Chamber of Commerce in
Rochester, New York, where he lived, is derived from the magic line
of the *lo-shu*. A typical example of a magic line is shown in Figure 66, where it is drawn on the Dürer square.

One of the great unsolved problems of recreational mathematics
is that of finding a method for calculating the number of different
squares of a given order. The number of order-5 squares was determined in 1973 by Rich Schroeppel. It is 275,305,224. The number of
order-6 squares is not yet known, but it is believed to have 20 digits.

The number of fifth-order diabolic squares was established by
Rosser and Walker as 28,800 (this includes rotations and reflections). Diabolic squares are possible in all orders above four except
those divisible by 2 but not by 4. There is none, for example, of
order six. Diabolic cubes and hypercubes also exist, but (as Rosser
and Walker have shown in unpublished papers) there are no cubes
of orders 3, 5, 7, $8k + 2$, $8k + 4$, or $8k + 6$, where k is any integer.
Diabolic cubes are possible in all other orders.

POSTSCRIPT

The worldwide literature on magic squares is now so vast that even a carefully pruned list of major references would require several pages. I shall be content with steering the reader to other books in this series, and to some references outside the series.

Varied kinds of order-3 magic squares are covered in two articles reprinted in *Gardner's Workout* (A K Peters, 2001). Some little known properties of the *lo-shu* are discussed in Chapter 21 of my Book 13.

Magic squares made with primes are covered in Book 5 (Chapter 9) and in Book 13 (Chapter 21). *Gardner's Workout* reprints an article on prime magic squares. It tells of my offer of $100 for a 3 × 3 magic square made with distinct primes in arithmetic progression. Harry Nelson won the prize with 22 computer-generated solutions. The one with the lowest constant is made with nine 10-digit integers.

The greatest unsolved question involving 3 × 3 magic squares is whether a square exists with nine distinct *square* numbers. I am still offering $100 for an example or a proof of impossibility. Such squares with integers raised to the power of n have been proved impossible for all n greater than 2. Only when $n = 2$ is the question open.

Lee Sallows's astonishing order-3 alphamagic square is featured in Chapter 21 of Book 13. At the left of Figure 67 is an order-3 magic square. Replace each of its numbers with a number that counts the letters in the English *name* of the number. The result is shown on the right. Incredibly, not only is it another magic square, but its integers

5	22	18
28	15	2
12	8	25

4	9	8
11	7	3
6	5	10

Figure 67. Lee Sallows's alphamagic squares.

6	8	9	7
3	12	5	11
10	1	14	13
16	15	4	2

Figure 68

are in consecutive order! Surely this is one of the most spectacular of all number coincidences.

Gardner's Workout also contains my review of a book (see the bibliography) that contains Dame Kathleen Ollerenshaw's brilliant solution to a difficult long-standing problem involving perfect pandiagonal squares.

Magic cubes are the topic of Chapter 17 in Book 12. Magic stars and magic polyhedrons are discussed in Chapter 5 of Book 6 and Chapter 17 of Book 9. An article on magic hexagrams is reprinted in my *Are Universes Thicker Than Blackberries?* (Norton, 2003).

Chapter 2 of Book 9 introduces antimagic squares of order-3 in which no two of the nine sums are alike. Squares made with dominoes are in Book 8 (Chapter 12), and with playing cards in Book 10 (Chapter 8).

Much work has been done on knight tours of magic squares. They are discussed in Book 7 (Chapter 14). There are many order-8 (chessboard) semi-magic squares that permit a knight to start on 1 and tour the chessboard by jumping to the cells in numerical order, but such squares are not magic along main diagonals. Not until 2003 did a lengthy computer search prove that no knight tours are possible on a fully magic 8×8 square.

Donald Knuth tells me that the magic square shown in Figure 63 is no longer the earliest known fourth-order square. Such squares appear in much earlier Islamic literature, as do magic squares of larger size. He cites as a source a French book by Sesiano, *Les carr ees magiques dans lespays islamiques* (2004).

Are antimagic squares in which the different sums are in arithmetic progression possible? Such a square of order-3 is impossible, but they can be constructed for higher orders. Clifford Pickover, in his book *The Zen of Magic Squares, Circles, and Stars* (see bibliography), gives (page 110) the following order-4 example shown in Figure 68.

On page 200 Pickover publishes Harvey Heinz's discovery of antimagic squares of orders 4 through 9 in which all sums are in arithmetic progression. Such constructions are easier, he writes, the larger the square, but there is as yet no procedure for making them.

BIBLIOGRAPHY

Magic Squares and Cubes. W. S. Andrews. The Open Court Publishing Company, 1917. Reprinted by Dover Publications in 1960.

"Magic Squares Made with Prime Numbers." W. S. Andrews and Harry A. Sayles in *The Monist* 23 (October 1918): 623–639.

"Magic Lines in Magic Squares." Claude Bragdon in *The Frozen Fountain*, pages 74–85. Alfred A. Knopf, 1932.

"On the Transformation Group for Diabolic Magic Squares of Order Four." Barkley Rosser and R. J. Walker in *Bulletin of the American Mathematical Society* 44:6 (June 1938): 416–420.

"The Algebraic Theory of Diabolic Magic Squares." Barkley Rosser and R. J. Walker in *Duke Mathematical Journal* 5:4 (December 1939): 705–728.

"Melencolia I." Erwin Panofsky in *Albrecht Dürer*, Vol. 1, pages 156–171. Princeton University Press, 1943.

Magic Squares. John Lee Fultz. Open Court, 1974.

New Recreations with Magic Squares. William A. Benson and Oswald Jacoby. Dover, 1976.

Researches in Magic Squares. Akira Hirayama and Gakuho Abe. Osaka Kyoikutusho, Osaka, Japan, 1983.

"Alphamagic Squares." Lee Sallows in *Abacus* 4 (1986): 28–45 and *Abacus* 5 (1987): 20–29, 43.

"Magic Squares and Cubes." Martin Gardner in *Time Travel and Other Mathematical Bewilderments*. W. H. Freeman, 1988.

"Unsolved Problems on Magic Squares." Gakuho Abe in *Discrete Mathematics* 127 (1994): 3–13.

"Magic Square of Squares." John Robertson. *Mathematics Magazine* 69 (October 1996): 289–293.

Most-Perfect Pandiagonal Magic Squares. Dame Kathleen Ollerenshaw and David Bree. Institute of Mathematics and Its Applications, 1998.

Inlaid Magic Squares and Cubes. John Hendricks. Privately published, 1999.

"A New Hypothesis on Dürer's Enigmatic Polyhedron in His Copper Engraving Melencholia." P. Schreiber in *Historia Mathematica* 26 (1999): 369–377.

"Most-Perfect Magic Squares." Ian Stewart in *Scientific American* (November 1999). A summary of Dame Ollerenshaw's work.

"3 × 3 Magic Squares." Martin Gardner, in *Gardner's Workout*, A K Peters, 2001.

"Some New Discoveries About Magic Squares." Martin Gardner in *Gardner's Workout*, A K Peters, 2001.

Legacy of the Luoshu. Frank J. Swetz. Open Court, 2001.

The Zen of Magic Squares, Circles, and Stars. Clifford Pickover. Princeton, 2002. This marvelous work is crammed with material not available in any other book.

"Multimagic Squares." Harm Derksen et al. in *American Mathematical Monthly* 114 (October 2007).

Benjamin Franklin's Numbers. Paul C. Pasles. Princeton University Press, 2007.

CHAPTER THIRTEEN

James Hugh Riley Shows, Inc.

THE JAMES HUGH RILEY SHOWS, INC., is one of the country's largest nonexistent carnivals. When I heard it had opened at the edge of town, I drove out to the lot to see my old friend Jim Riley; we had been classmates some 20 years ago at the University of Chicago. Riley was then taking graduate courses in mathematics, but one summer he joined a carnival as a "talker" for the girlie show, and during most of the subsequent years he had been, as the carnies say, "with it." To everyone on the lot he was known simply as The Professor. Somehow he had managed to keep alive his passion for mathematics, and whenever we got together I could always count on picking up some unusual items for this department.

I found The Professor chatting with the ticket collector in front of the freak show. He was wearing a white Stetson hat and seemed older and heavier than when I had last seen him. "Read your column every month," he said as we pumped hands. "Ever thought about writing up Spot-the-spot?"

"Come again?" I said.

"It's one of the oldest games on the lot." He grabbed my arm and pushed me down the midway until we came to a concession where a red circular spot a yard in diameter was painted on the counter. The object of the game was to place five metal disks one at a time on this spot in such a way that they completely covered the spot. Each disk was about 22 inches across. Once a disk had been placed the player was not permitted to move it, and the game was lost if even the tiniest bit of red remained visible after the fifth disk was down.

"Of course," said The Professor, "we use the largest possible spot that can still be covered by the disks. Most people think the disks should go like so." He arranged them symmetrically on the spot as

Figure 69. An inferior method of placing the disks in Spot-the-spot. (Artist: Alex Semenoick)

shown in Figure 69. The circumference of each disk touched the spot's center, and the centers of the disks formed the corners of a regular pentagon. Five minute areas of red were visible around the spot's rim.

"Unfortunately," Riley continued, "that doesn't quite do it. To cover the maximum circular area, you have to arrange them this way." He pushed the disks with his finger until they assumed the formation shown in Figure 70. Disk 1, he explained, has its center on diameter AD and its circumference on point C, which is slightly below the spot's center (B). Disks 3 and 4 are then placed so their edges pass through C and D. Disks 2 and 5 cover the rest of the spot as shown.

Naturally I wanted to know the distance of BC. Riley couldn't remember exactly, but he later sent me the reference to an article in which this difficult problem is worked out in detail: "On the Solution of Numerical Functional Equations, Illustrated by an Account of a Popular Puzzle and Its Solution," by Eric H. Neville (*Proceedings of the London Mathematical Society*, Second Series, Vol. 14, pages 308–326; 1915). If the radius of the spot is 1, the distance BC is a trifle more than 0.0285 and the smallest radius possible for the disks is

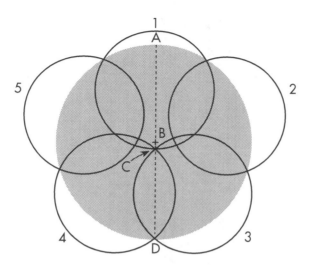

Figure 70. The correct method of placing the disks in Spot-the-spot. (Artist: Alex Semenoick)

0.609+. If the disks are placed as shown in Figure 69, they must have a radius of 0.6180339+ in order to cover the spot completely. (This number is the reciprocal of phi, the golden ratio discussed in Chapter 8.) The curious feature of the problem is the smallness of difference between the areas covered by the two methods of arranging the disks. Unless the spot is about a yard in diameter, the difference is scarcely detectable.

"This reminds me," said I, "of a fascinating minimal-area problem still unsolved. You define the diameter of an area as the longest straight line that will join two points on it. The question is: What are the shape and area of the smallest plane figure that will cover any area of unit diameter?"

The Professor nodded. "The smallest regular polygon that does it is a hexagon with a side of $1/\sqrt{3}$, but about 30 years ago someone improved this by chopping off two corners." He took a pencil and pad of paper from his jacket and sketched the pattern reproduced in Figure 71. The corners are sliced off along lines tangent to the inscribed circle (which has a unit diameter) and perpendicular to lines from the circle's center to the corners.

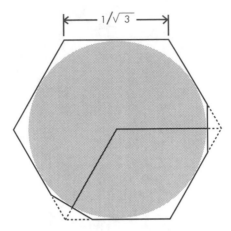

Figure 71. A truncated hexagon that will cover any area with a "diameter" of 1. (Artist: Alex Semenoick)

"Is that the best solution so far?" I asked.

Riley shook his head. "I've heard that a few years ago someone at the University of Illinois sliced off another small piece, but I don't know the details."

We sauntered down the midway and stopped in front of a concession where three enormous dice were tumbling down a corrugated incline to a flat surface below. Large white digits from 1 to 6 were painted on the counter. A player could put as much money as he wished on any digit. The dice were rolled. If his number appeared once on the dice, he received back his bet plus the same amount of money. If the number appeared twice, he got back his bet plus twice the amount. If the number showed on all three dice, he got back his bet plus three times the amount. Of course if the number did not show at all, he lost his bet.

"How can this game show a profit?" I asked. "The probability of a certain number showing on one die is 1/6, so with three dice the probability is 3/6 or 1/2 that the number will show at least once. If the number shows more than once, the player can win even more than he bets, so it looks to me like the game favors the player."

The Professor chuckled. "That's just how we want the marks [carny slang for suckers] to figure it. Think about it again." When I did think about it later, I was astonished. Perhaps some readers will

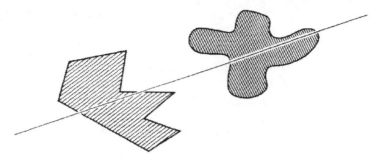

Figure 72. The "sandwich theorem" in two dimensions. (Artist: Alex Semenoick)

enjoy calculating just how much, in the long run, a player can expect to win for every dollar that he bets.

Before I left the lot, Riley took me into one of his "grab joints" (as he called them) for a bite to eat. Our coffee was served at once, but I decided not to touch it until our sandwiches came.

"If you want to keep your coffee hot," The Professor said, "better pour your cream now instead of later. The hotter the coffee, the faster its rate of heat loss."

I dutifully poured my cream.

When The Professor's ham sandwich arrived, sliced neatly through the middle, he gazed at it for a moment and said, "Have you ever come across a paper by Tukey and Stone on the generalized ham-sandwich theorem?"

"You mean John Tukey and Arthur Stone? Two of the co-discoverers of flexagons?"

"The same."

I shook my head. "I don't even know about the ungeneralized ham-sandwich theorem."

Riley took out his pad again and drew a line segment on it. "Any one-dimensional figure can always be bisected by one point. Right?" I nodded while he drew two irregular closed curves, then a straight line that sliced both of them (see Figure 72). "Any pair of areas on a plane can be exactly bisected by one straight line. Correct?"

"I'll take your word for it."

"It's not hard to prove. There's an elementary proof in *What Is Mathematics?* by Richard Courant and Herbert Robbins. It makes use of Bolzano's theorem."

"Ah, yes," I said. "If a continuous function of x has positive and negative values, it has to have at least one zero value."

"Right. It seems trivial, but it's a powerful tool in all sorts of existence proofs. Of course in this case, the proof doesn't tell you how to construct the line. It only proves that the line exists."

"Where do ham sandwiches come in?"

"When we move on to three dimensions. The volumes of any three solids, of any size or shape, placed anywhere in space, can always be exactly and simultaneously bisected by a plane – like bisecting two pieces of bread and a slice of ham in between. Stone and Tukey generalized this for all dimensions. They proved that there is always a hyperplane that bisects four four-dimensional solids placed anywhere in four-dimensional space, or five five-dimensional solids, and so on."

The Professor drained his cup, then pointed across the counter to a pile of doughnuts. "Speaking of slicing solids, here's a curious question you might ask your readers sometime. What's the maximum number of pieces you can get with three simultaneous plane cuts through one doughnut? It's a problem I thought of myself."

I closed my eyes and tried to visualize it while the merry-go-round calliope wheezed off key, but the problem made my head throb and I finally gave up.

ADDENDUM

The carnival game with the three dice is known in the United States as Chuck-a-luck or Bird Cage. It is a popular dice game in gambling casinos, where the dice are tumbled inside a wire cage called the chuck cage. It is sometimes gaffed with electromagnets (see *Scarne on Dice*, by John Scarne and Clayton Rawson, Military Service Publishing Company, 1945, pages 333–335). The game is also discussed in Chapter 7 of *Facts from Figures*, a Penguin paperback by M. J. Moroney. Moroney calls it the Crown and Anchor game because in England it is often played with dice bearing hearts, clubs, spaces, diamonds, crowns, and anchors.

"The game is beautifully designed," Moroney writes. "In over half the throws the banker sees nothing for himself. Whenever he makes a profit, he pays out more bountifully to other people, so that the losers' eyes turn to the lucky winner in envy, rather than to the banker in suspicion. Spectacular wins are kept to the minimum, but when they do fall the blow is always softened by apparent generosity."

A number of readers took issue with The Professor's suggestion that it is best to pour cream immediately in order to conserve the heat of a cup of coffee. Unfortunately, these readers were about equally divided between those who thought heat was best conserved by pouring the cream later and those who thought it made no difference when the cream was poured.

I asked Norman T. Gridgeman, a statistician with the National Research Council of Canada, in Ottawa, to look into the matter and I am happy to say that his analysis confirms The Professor's statement. On the basis of Newton's law of cooling (which states that the rate of heat loss is proportional to the difference between the temperature of the hot material and the temperature of the ambient), and taking into consideration the significant and easily overlooked fact that the volume of the coffee increases after the cream is added, it turns out that an immediate mixing of the liquids always conserves heat. This is true regardless of whether the cream is at ambient temperature or below. Other factors such as changes in the rate of radiation due to the lightened color of the liquid, an increased surface area in cups with sloping sides, and so on, have a negligible influence.

A typical example is as follows. The initial temperature of 250 grams of coffee is 90 degrees, the initial temperature of 50 grams of cream is 10 degrees, and the ambient is 20 degrees. If the cream is added immediately, the temperature of the coffee 30 minutes later will be about 48 degrees. If the cream is not added until after 30 minutes have elapsed, the resulting temperature will be about 45 degrees – a difference of 3 degrees.

ANSWERS

A person playing the carnival dice game can expect to win a trifle more than 92 cents for every dollar bet. There are 216 equally probable ways three dice can fall, of which 91 are wins for the player. His chances of winning something on each bet, therefore, are 91/216. Assume that he plays the game 216 times, betting one dollar each time, and that each time the three dice fall a different way. On 75 of his wins his number appears only once, so he is paid $150 by the operator. On 15 wins the number shows twice, so he is paid $45. On one win all three dice will show the number, earning him $4. The total paid to him is $199. To win this, he bet $216; consequently, he expects in the long run, for every dollar wagered, to win 199/216 dollars, or $0.9212+. This gives a little more than 7.8 cents to the operator on every dollar bet: a profit of about 7.8 percent.

Figure 73 shows how a doughnut can be sliced into 13 pieces by three simultaneous plane cuts. A large number of correspondents sent correct solutions, but a majority failed to find that elusive thirteenth piece. The formula for the largest number of pieces that can be produced with n cuts is

$$\frac{n^3 + 3n^2 + 8n}{6}$$

If one is permitted to rearrange the pieces after each cut, as many as 18 pieces can be obtained.

Many interesting letters about the doughnut-slicing problem were received. Derrill Bordelon, of the U.S. Naval Ordnance Laboratory at Silver Spring, Maryland, sent a detailed proof of the formula for n cuts. Dan Massey, Jr., of Chattanooga, Tennessee, speculated on a formula for n-dimensional doughnuts. Richard Gould, Menlo Park, California, wrote in the margin of a letter than he had obtained such a generalized formula but that the margin was too small to contain it. John McClellan, Woodstock, New York, raised the difficult question: What is the optimum proportion of the diameter of the doughnut's hole to the diameter of its cross section, in order to obtain the largest possible *smallest* piece?

David B. Hall, Towson, Maryland, after some careful tests with actual doughnuts, wrote

Sirs:
    A little study of the problem indicated that there should be a maximum of thirteen pieces. This would have closed the matter, except that the next time I was at the grocer's I bought a box of doughnuts

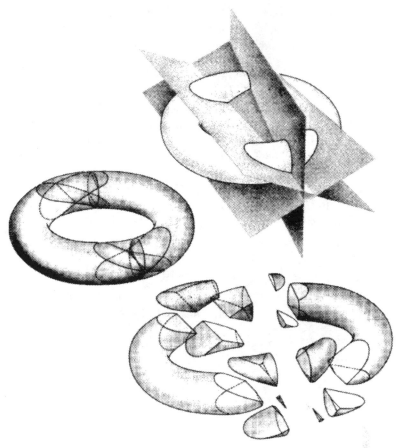

Figure 73. How to slice a doughnut into 13 pieces with only three plane cuts. (Artist: James Egleson)

and discovered that the technical problems were as intriguing as the mathematical one.

Obtaining thirteen pieces involves carving out a slender pyramid with its vertex embedded in the body of the doughnut. After finding that reasonably predictable cuts could be made with embedded toothpicks as guides, I made my first full-scale section, only to discover that no trace of the two smallest pyramids could be found. (There were plenty of crumbs, but I suppose they don't count.) It turns out that the requirement that three planes be cut through a doughnut necessitates not only care in cutting but very thorough provision against movement of wedge-shaped pieces under pressure as successive cuts are made. In this case the parts containing the tiny pyramids had spread very slightly, but enough to escape the knife completely.

On my final doughnut, using steel skewers instead of toothpicks, I achieved complete success and obtained fifteen well-defined pieces. The pyramids were more than successful. By overzealously preventing the previous spreading I was able to get a little overlap instead. The two bonus pieces resulted from the fact that the hole was not very round and each of the first two cuts yielded a small but honest knob.

A very thin hula-hoop-shaped doughnut might make cutting easier, but this arrangement was discovered after the doughnuts were eaten and has not been explored.

BIBLIOGRAPHY

"Generalized 'Sandwich' Theorems." A. H. Stone and J. W. Tukey in *Duke Mathematical Journal* 9: 2 (June 1942): 356–359.

CHAPTER FOURTEEN

Nine More Problems

1. CROSSING THE DESERT

An unlimited supply of gasoline is available at one edge of a desert 800 miles wide, but there is no source on the desert itself. A truck can carry enough gasoline to go 500 miles (this will be called one "load"), and it can build up its own refueling stations at any spot along the way. These caches may be any size, and it is assumed that there is no evaporation loss.

What is the minimum amount (in loads) of gasoline the truck will require in order to cross the desert? Is there a limit to the width of a desert the truck can cross?

2. THE TWO CHILDREN

Mr. Smith has two children. At least one of them is a boy. What is the probability that both children are boys?

Mr. Jones has two children. The older child is a girl. What is the probability that both children are girls?

3. LORD DUNSANY'S CHESS PROBLEM

Admirers of the Irish writer Lord Dunsany do not need to be told that he was fond of chess. (Surely his story "The Three Sailors' Gambit" is the funniest chess fantasy ever written.) Not generally known is the fact that he liked to invent bizarre chess problems which, like his fiction, combine humor and fantasy.

The problem depicted in Figure 74 was contributed by Dunsany to *The Week-End Problems Book*, compiled by Hubert Phillips,

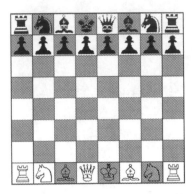

Figure 74. Lord Dunsany's chess problem.

Nonesuch Press, 1932. Its solution calls more for logical thought than skill at chess, although one does have to know the rules of the game. White is to play and mate in four moves. The position is one that could occur in actual play.

4. PROFESSOR ON THE ESCALATOR

When Professor Stanislaw Slapenarski, the Polish mathematician, walked very slowly down the down-moving escalator, he reached the bottom after taking 50 steps. As an experiment, he then ran up the same escalator, one step at a time, reaching the top after taking 125 steps.

Assuming that the professor went up five times as fast as he went down (e.g., took five steps to every one step before), and that he made each trip at a constant speed, how many steps would be visible if the escalator stopped running?

5. THE LONESOME 8

The most popular problem ever published in *The American Mathematical Monthly*, its editors recently disclosed, is the following. It was contributed by P. L. Chessin of the Westinghouse Electric Corporation to the April 1954 issue.

"Our good friend and eminent numerologist, Professor Euclide Paracelso Bombasto Umbugio, has been busily engaged in testing on his desk calculator the 81×10^9 possible solutions to the problem of reconstructing the following exact long division in which the digits were indiscriminately replaced by X save in the quotient where they were almost entirely omitted:

$$
\begin{array}{r}
8 \\
\text{XXX)}\overline{\text{XXXXXXXX}} \\
\underline{\text{XXX}} \\
\text{XXXX} \\
\underline{\text{XXX}} \\
\text{XXXX} \\
\underline{\text{XXXX}}
\end{array}
$$

"Deflate the Professor! That is, reduce the possibilities to $\left(81 \times 10^9\right)^0$."

Because any number raised to the power of zero is one, the reader's task is to discover the unique reconstruction of the problem. The 8 is in correct position above the line, making it the third digit of a five-digit answer. The problem is easier than it looks, yielding readily to a few elementary insights.

6. DIVIDING THE CAKE

There is a simple procedure by which two people can divide a cake so that each is satisfied he has at least half: One cuts and the other chooses. Devise a general procedure so that n persons can cut a cake into n portions in such a way that everyone is satisfied he has at least $1/n$ of the cake.

7. THE FOLDED SHEET

Mathematicians have not yet succeeded in finding a formula for the number of different ways a road map can be folded, given n creases in the paper. Some notion of the complexity of this question can be gained from the following puzzle invented by the British puzzle expert Henry Ernest Dudeney.

Divide a rectangular sheet of paper into eight squares and number them on one side only, as shown at top left in Figure 75. There are 40 different ways that this "map" can be folded along the ruled lines to form a square packet that has the "1" square face-up on top

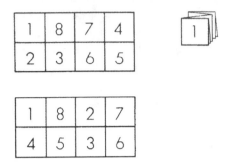

Figure 75. Dudeney's map-folding puzzle. (Artist: Alex Semenoick)

and all other squares beneath. The problem is to fold this sheet so that the squares are in serial order from 1 to 8, with the 1 face-up on top.

If you succeed in doing this, try the much more difficult task of doing the same thing with the sheet numbered in the manner pictured at the bottom of the illustration.

8. THE ABSENT-MINDED TELLER

An absent-minded bank teller switched the dollars and cents when he cashed a check for Mr. Brown, giving him dollars instead of cents, and cents instead of dollars. After buying a five-cent newspaper, Brown discovered that he had left exactly twice as much as his original check. What was the amount of the check?

9. WATER AND WINE

A familiar chestnut concerns two beakers, one containing water, the other wine. A certain amount of water is transferred to the wine, and then the same amount of the mixture is transferred back to the water. Is there now more water in the wine than there is wine in the water? The answer is that the two quantities are the same.

Raymond Smullyan writes to raise the further question: Assume that at the outset one beaker holds 10 ounces of water and the other holds 10 ounces of wine. By transferring three ounces back and forth any number of times, stirring after each transfer, is it possible to reach a point at which the percentage of wine in each mixture is the same?

ANSWERS

1. The following analysis of the desert-crossing problem appeared in a recent issue of *Eureka*, a publication of mathematics students at the University of Cambridge. Five hundred miles will be called a "unit"; gasoline sufficient to take the truck 500 miles will be called a "load"; and a "trip" is a journey of the truck in either direction from one stopping point to the next.

   Two loads will carry the truck a maximum distance of 1 and 1/3 units. This is done in four trips by first setting up a cache at a spot 1/3 unit from the start. The truck begins with a full load, goes to the cache, leaves 1/3 load, returns, picks up another full load, arrives at the cache and picks up the cache's 1/3 load. It now has a full load, sufficient to take it the remaining distance to one unit.

   Three loads will carry the truck 1 and 1/3 plus 1/5 units in a total of nine trips. The first cache is 1/5 unit from the start. Three trips put 6/5 loads in the cache. The truck returns, picks up the remaining full load, and arrives at the first cache with 4/5 load in its tank. This, together with the fuel in the cache, makes two full loads, sufficient to carry the truck the remaining 1 and 1/3 units, as explained in the preceding paragraph.

   We are asked for the minimum amount of fuel required to take the truck 800 miles. Three loads will take it 766 and 2/3 miles (1 and 1/3 plus 1/5 units), so we need a third cache at a distance of 33 and 1/3 miles (1/15 unit) from the start. In five trips the truck can build up this cache so that when the truck reaches the cache at the end of the seventh trip, the combined fuel of truck and cache will be three loads. As we have seen, this is sufficient to take the truck the remaining distance of 766 and 2/3 miles. Seven trips are made between starting point and first cache, using 7/15 load of gasoline. The three loads of fuel that remain are just sufficient for the rest of the way, so the total amount of gasoline consumed will be 3 and 7/15, or a little more than 3.46 loads. Sixteen trips are required.

Proceeding along similar lines, four loads will take the truck a distance of 1 and 1/3 plus 1/5 plus 1/7 units, with three caches located at the boundaries of these distances. The sum of this infinite series diverges as the number of loads increases; therefore, the truck can cross a desert of any width. If the desert is 1,000 miles across, seven caches, 64 trips, and 7.673 loads of gasoline are required.

Hundreds of letters were received on this problem, giving general solutions and interesting sidelights. Cecil G. Phipps, professor of mathematics at the University of Florida, summed matters up succinctly as follows:

The general solution is given by the formula:

$$d = m(1 + 1/3 + 1/5 + 1/7 + \cdots)$$

where d is the distance to be traversed and m is the number of miles per load of gasoline. The number of depots to be established is one less than the number of terms in the series needed to exceed the value of d. One load of gasoline is used in the travel between each pair of stations. Since the series is divergent, any distance can be reached by this method although the amount of gasoline needed increases exponentially.

If the truck is to return eventually to its home station, the formula becomes

$$d = m(1/2 + 1/4 + 1/6 + 1/8 + \cdots)$$

This series is also divergent and the solution has properties similar to those for the one-way trip.

Many readers called attention to three previously published discussions of the problem:

Problem in Logistics: The Jeep Problem. Olaf Helmer. Project Rand Report No. RA-15015, December 1, 1946. This was the first unclassified report of the Rand publication, issued when the project was still under the wing of Douglas Aircraft Company. It is the clearest analysis of the problem, including the return-trip version, that I have seen.

"Crossing the Desert." G. G. Alway in the *Mathematical Gazette* 41:337 (October 1947): 209.
"The Jeep Problem: A More General Solution." C. G. Phipps in the *American Mathematical Monthly* 54:8 (October 1947): 458–462.

2.   If Smith has two children, at least one of which is a boy, we have three equally probable cases:

> Boy-boy
> Boy-girl
> Girl-boy

In only one case are both children boys, so the probability that both are boys is 1/3.

Jones's situation is different. We are told that his older child is a girl. This limits us to only two equally probable cases:

> Girl-girl
> Girl-boy

Therefore the probability that both children are girls is 1/2.

[This is how I answered the problem in my column. After reading protests from many readers, and giving the matter considerable further thought, I realized that the problem was ambiguously stated and could not be answered without additional data. For a later discussion of the problem, see Chapter 19.]

3.   The key to Lord Dunsany's chess problem is the fact that the black queen is not on a black square as she must be at the start of a game. This means that the black king and queen have moved, and this could have happened only if some black pawns have moved. Pawns cannot move backward, so we are forced to conclude that the black pawns reached their present positions from the other side of the board! With this in mind, it is easy to discover that the white knight on the right has an easy mate in four moves.

White's first move is to jump his knight at the lower right corner of the board to the square just above his king. If black moves the upper left knight to the rook's file, white mates in two more moves. Black can, however, delay the mate one

move by first moving his knight to the bishop's file instead of
the rook's. White jumps his knight forward and right to the
bishop's file, threatening mate on the next move. Black moves
his knight forward to block the mate. White takes the knight
with his queen and then mates with his knight on the fourth
move.

4. Let n be the number of steps visible when the escalator is not
    moving, and let a unit of time be the time it takes Professor
    Slapenarski to walk down one step. If he walks down the down-
    moving escalator in 50 steps, then $n - 50$ steps have gone out
    of sight in 50 units of time. It takes him 125 steps to run up the
    same escalator, taking five steps to every one step before. In this
    trip, $125 - n$ steps have gone out of sight in 125/5, or 25, units
    of time. Since the escalator can be presumed to run at constant
    speed, we have the following linear equation that readily yields
    a value for n of 100 steps:

$$\frac{n-50}{50} = \frac{125 - n}{25}$$

5. In long division, when two digits are brought down instead of
    one, there must be a zero in the quotient. This occurs twice, so
    we know at once that the quotient is X080X. When the divisor is
    multiplied by the quotient's last digit, the product is a four-digit
    number. The quotient's last digit must therefore be 9 because
    eight times the divisor is a three-digit number.

    The divisor must be less than 125 because eight times 125 is
    1,000, a four-digit number. We now can deduce that the quo-
    tient's first digit must be more than 7, for seven times a divisor
    less than 125 would give a product that would leave more than
    two digits after it was subtracted from the first four digits in the
    dividend. This first digit cannot be 9 (which gives a four-digit
    number when the divisor is multiplied by it), so it must be 8,
    making the full quotient 80,809.

    The divisor must be more than 123 because 80,809 times 123
    is a seven-digit number and our dividend has eight digits. The
    only number between 123 and 125 is 124. We can now recon-
    struct the entire problem as follows:

$$\frac{80809}{124)\overline{10020316}}$$
$$\underline{992}$$
$$1003$$
$$\underline{992}$$
$$1116$$
$$\underline{1116}$$

6. Several procedures have been devised by which n persons can divide a cake in n pieces so that each is satisfied that he has at least $1/n$ of the cake. The following system has the merit of leaving no excess bits of cake.

Suppose there are five persons: A, B, C, D, E. A cuts off what he regards as $1/5$ of the cake and what he is content to keep as his share. B now has the privilege, if he thinks A's slice is more than $1/5$, of reducing it to what he thinks is $1/5$ by cutting off a portion. Of course if he thinks it is $1/5$ or less, he does not touch it. C, D, and E in turn now have the same privilege. The last person to touch the slice keeps it as his share. Anyone who thinks that this person got less than $1/5$ is naturally pleased because it means, in his eyes, that more than $4/5$ remains. The remainder of the cake, including any cut-off pieces, is now divided among the remaining four persons in the same manner, then among three. The final division is made by one person cutting and the other choosing. The procedure is clearly applicable to any number of persons.

For a discussion of this and other solutions, see the section "Games of Fair Division," pages 363–368, in *Games and Decisions*, by R. Duncan Luce and Howard Raiffa, John Wiley and Sons, 1957.

7. The first sheet is folded as follows. Hold it face down so that when you look down on it the numbered squares are in this position:

$$\frac{2365}{1874}$$

Fold the right half on the left so that 5 goes on 2, 6 on 3, 4 on 1, and 7 on 8. Fold the bottom half up so that 4 goes on 5, and 7 on 6. Now tuck 4 and 5 between 6 and 3, and fold 1 and 2 under the packet.

The second sheet is first folded in half the long way, the numbers outside, and held so that 4536 is uppermost. Fold 4 on 5. The right end of the strip (squares 6 and 7) is pushed between 1 and 4, then bent around the folded edge of 4 so that 6 and 7 go between 8 and 5, and 3 and 2 go between 1 and 4.

8. To determine the value of Brown's check, let x stand for the dollars and y for the cents. The problem can now be expressed by the following equation: $100y + x - 5 = 2(100x + y)$. This reduces to $98y - 199x = 5$, a Diophantine equation with an infinite number of integral solutions. A solution by the standard method of continued fractions gives as the lowest values in positive integers: $x = 31$ and $y = 63$, making Brown's check \$31.63. This is a unique answer to the problem because the next lowest values are: $x = 129$, $y = 262$, which fails to meet the requirement that y be less than 100.

There is a much simpler approach to the problem and many readers wrote to tell me about it. As before, let x stand for the dollars on the check, y for the cents. After buying his newspaper, Brown has left $2x + 2y$. The change that he has left, from the x cents given him by the cashier, will be $x - 5$.

We know that y is less than 100, but we don't know yet whether it is less than 50 cents. If it is less than 50 cents, we can write the following equations:

$$2x = y$$
$$2y = x - 5$$

If y is 50 cents or more, then Brown will be left with an amount of cents ($2y$) that is a dollar or more. We therefore have to modify the above equations by taking 100 from $2y$ and adding 1 to $2x$. The equations become

$$2x + 1 = y$$
$$2y - 100 = x - 5$$

Each set of simultaneous equations is easily solved. The first set gives x a minus value, which is ruled out. The second set gives the correct values.

9. Regardless of how much wine is in one beaker and how much water is in the other, and regardless of how much liquid is transferred back and forth at each step (provided it is not all of the liquid in one beaker), it is impossible to reach a point at which the percentage of wine in each mixture is the same. This can be shown by a simple inductive argument. If beaker A contains a higher concentration of wine than beaker B, then a transfer from A to B will leave A with the higher concentration. Similarly a transfer from B to A – from a weaker to a stronger mixture – is sure to leave B weaker. Since every transfer is one of these two cases, it follows that beaker A must always contain a mixture with a higher percentage of wine than B. The only way to equalize the concentrations is by pouring all of one beaker into the other.

There is a fallacy in the above solution. It assumes that liquids are infinitely divisible, whereas they are composed of discrete molecules. P. E. Argyle of Royal Oak, British Columbia, set me straight with the following letter:

> Sirs:
> Your solution to the problem of mixing wine and water seems to ignore the physical nature of the objects involved. When a sample of fluid is taken from a mixture of two fluids, the proportion of one fluid present in the sample will be different from its proportion in the mixture. The departure from the "correct" amount will be of the order $\pm\sqrt{n}$, where n is the number of molecules expected to be present.
>
> Consequently it is possible to have equal amounts of wine in the two glasses. The probability of this occurring becomes significant after the expected lack of equality in the mixture has been reduced to the order of $\sqrt{n}$. This requires only 47 double interchanges for the problem as it was stated.

POSTSCRIPT

The cake-cutting problem obviously has many applications. A married couple, for example, wants to divide household chores fairly between them. Or three or more people who share a house want to do the same.

In the procedure I gave for *n* persons, each person is satisfied he or she got a fair share. Is there a procedure that in addition guarantees that each person is persuaded that *everyone else* is similarly convinced he or she got a fair share? This stronger version has been the topic of many technical papers.

A good reference in addition to the one I gave is "How to Cut a Cake Fairly," by L. E. Dubins and E. H. Spanker, in *American Mathematical Monthly* 68 (January 1961): 1–17. In 1998, A K Peters published an entire book on the problem: *Cake Cutting Algorithms: Be Fair if You Can*, by Jack Robertson and William Webb.

The traditional problem of the water and wine is discussed in Chapter 10 of Book 1. There it is explained how the problem can be beautifully modeled by a surprising trick with playing cards.

Other references include the following:

"An Energy-Free Cake Division Protocol." Steven J. Brams and Alan Taylor in *American Mathematical Monthly* 102 (1995): 9–19.

Fair Division: From Cake Cutting to Dispute Resolution. Steven J. Brams and Alan Taylor. Cambridge University Press, 1996.

The Win-Win Solution: Guaranteeing Fair Shares to Everybody. Steven J. Brams and Alan Taylor. Norton, 1999.

"Toward a Fairer Expansion Draft." Ivars Peterson in *Mathematical Treks*. Mathematical Association of America, 2002.

"Cake-Cutting." David Darling in *The Universal Book of Mathematics*. Castle Books, 2004.

"Better Ways to Cut a Cake." Steven J. Brams et al. in *Notices of the AMS* 53 (December 2006): 1314–1321. It lists 25 references.

How To Cut a Cake and Other Mathematical Conundrums. Ian Stewart. Oxford, 2006.

CHAPTER FIFTEEN

Eleusis: The Induction Game

MOST MATHEMATICAL GAMES, from ticktacktoe to chess, call for deductive reasoning on the part of the players. In contrast, Eleusis, a remarkable new card game devised by Robert Abbott, is an induction game. Abbott is a young New York writer who has invented a large number of offbeat card and board games, but this one is of special interest to mathematicians and other scientists because of its striking analogy with scientific method and its exercise of precisely those psychological abilities in concept formation that seem to underlie the "hunches" of creative thinkers.

Eleusis (pronounced ee-loo-sis) is a game for three or more players. It makes use of the standard deck of playing cards. Players take turns at being the "dealer," who has no part in the actual play except to serve as a kind of umpire. He deals to the other players until one card remains. This is placed face up in the center of the table as the first card of the "starter pile." To make sure that players receive equal hands, the dealer must remove a certain number of cards before dealing. For three players (including the dealer, who of course does not get a hand), he removes one card; for four players, no cards; five players, three cards, and so on. The removed cards are set aside without being shown.

After the cards are dealt and the "starter card" is in place, the dealer makes up a secret rule that determines what cards can be played on the starter pile. It is this rule that corresponds to a law of science; the players may think of the deal as Nature, or, if they prefer, as God. The dealer writes his rule on a piece of paper, which he folds and puts aside. This is for later checking to make sure that the dealer does not upset Nature's uniformity by changing his rule. For each player the object of the game is to get rid of as many cards as possible. This can be done rapidly by any player who correctly guesses the secret rule.

An example of a very simple rule is: "If the top card of the starter pile is red, play a black card. If the top card is black, play a red card." Beginners should limit themselves to extremely simple rules of this type, and then move on to more complicated rules as their ability to play improves. One of the most ingenious features of Eleusis is that the method of scoring (to be explained later) puts pressure on the dealer to choose a rule that not everyone will guess quickly, but that is simple enough so that one player is likely to guess it ahead of the others and fairly early in the game. Here again we have a pleasant analogy. The basic laws of physics are difficult to detect, yet once they are discovered they usually turn out to be based on relatively simple equations.

After the rule is written, the "first stage" of the game begins. The first player takes any card from his hand and places it face up on the starter card. If the card conforms to the secret rule, the dealer says "Right" and the card remains on the starter pile. If it violates the rule, the dealer says "Wrong." The player then takes back the card, places it face up in front of him, and the turn passes to the next player on the left. Each player must play one card from his hand at each turn. His "mistake cards" are left face up in front of him and spread slightly so that they can be clearly identified. The correctly played cards which form the starter pile are also fanned along the table so that all the cards can be seen. A typical starter pile is shown in Figure 76.

Each player tries to analyze the cards in the starter pile to discover the rule governing their sequence. He then forms a hypothesis that he can test by playing what he thinks is a correct card, or by playing a card he suspects will be rejected. The first stage of the game ends when all the cards in the players' hands have been played.

The dealer's score is now figured. It is based on how far the leading player (the person with the fewest mistake cards) is ahead of the others. If there are two players (not counting the dealer), the dealer's score is the number of cards in the leading player's mistake pile subtracted from the number of cards in the other player's pile. For three players, multiply the leading player's mistake cards by two, then subtract from the total of mistake cards belonging to the other

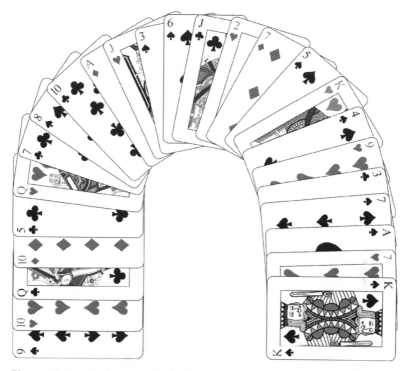

Figure 76. A typical starter pile for the game of Eleusis. What is the secret rule that determines the order of the cards? (Artist: Harold Jacobs)

players. For four players, multiply by three and do the same. For five players, the multiplier is four; for six the multiplier is five, and so on. The suits and values of cards do not enter into the scoring.

For example, suppose there are three players and the dealer. The mistake cards number 10, 5, and 3. Twice 3 is 6, which is taken from 15 to give the dealer a score of 9. This is recorded and the game goes into its second and final stage, during which the mistake cards are played.

The mistake cards remain fanned face up on the table in front of each player, but a player may rearrange his cards if he wishes. Plays are made in turn as before, each player taking any card and putting it on the starter pile. The dealer tells him if it is right or wrong. If it is wrong, he replaces the card among his mistake cards. The second stage ends when one player gets rid of all his cards, or when the dealer sees that it is impossible for more cards to be accepted on the starter pile.

The slip of paper is now opened and the rule read. This corresponds in a sense to the mathematician's final deductive proof of a theorem that was first suggested to him by an inductive guess based on a set of particular observations. Scientists are of course denied this final verification and must rest content with establishing their hypotheses to a high degree of probability. If the scientist accepts the pragmatic epistemology of, say William James and John Dewey, he may not believe in the existence of the folded sheet of paper. The successful operation of his hypothesis will be the only meaning of its "truth." Or he may agree with Bertrand Russell and others that the truth of his theory is its correspondence with an external structure, even though he has no way of seizing the structure and unfolding it. Still another point of view is favored by Rudolf Carnap and his friends. To ask if there "exists" a folded slip of paper (i.e., a final structure of some sort to which scientific theories correspond) is to ask a pseudo-question. Since there is no way such a question can be answered, it should be replaced by the practical question: Given a certain context for discourse, what is the best language form to use when talking about scientific laws and theories?

Players are now scored in a manner similar to the way in which the dealer was scored. Each takes the number of cards he holds, multiplies by the number of players exclusive of himself and the dealer, and then subtracts the product from the total number of cards held by the other players. If the result is a minus number, he is given a score of 0. A bonus of 6 goes to the player who went out. If no one went out, it goes to the player with the fewest cards, and if two or more tie, the bonus is divided between them. For example, if there are four players (excluding the dealer) who hold 2, 3, 10, and 0 cards, their respective scores will be 7, 3, 0, and 21.

The deal passes to the left after each hand. The game continues until each person has been dealer twice; then the player with the highest score is the winner of the set.

If the rule is not applicable until two cards are on the starter pile, then the first card played is correct no matter what it is. If a rule involves numbers, the ace is 1, the jack 11, the queen 12, and the king 13. If it is permissible to "turn the corner" (continue in cyclic fashion: J-Q-K-A-2–3 . . .), the dealer must state this in his rule.

Rules that restrict a player, on most of his turns, to fewer than a fifth of the cards in the deck should be avoided. For example, the rule "Play a card with a value of one unit above the value of the top card" is not acceptable, because at each turn a player would be limited to only four cards out of the 52.

After writing down his rule, the dealer may, if he wishes, give a hint of it. He might say: "This rule involves the two top cards of the starter pile," or "This rule involves the suits." After the play begins, no further hints are permitted unless the play is very informal.

The following secret rules are typical, and are listed in order of increasing complexity.

1. Alternate even and odd cards.
2. The card played must have either the same suit or the same value as the card on top of the pile (as in the card game called Eights).
3. If the top two cards are of the same color, play a card from ace to 7. If they are of different colors, play a card from 7 to king.
4. If the second card from the top is red, play a card with a value equal to or higher than this card. If the second card is black, play a card of equal or lower value.
5. Divide the value of the top card by four. If the remainder is one, play a spade; if two, play a heart; if three, play a diamond; if zero, play a club.

If the players have some mathematical sophistication, the rules can of course be more advanced. The dealer, however, must always shrewdly estimate the skill of the players so that he can raise his score by choosing a rule that one player is likely to discover ahead of the others.

It is permissible to make up rules in which the players themselves are involved. (One thinks of the physicist whose apparatus influences what he is trying to observe or the anthropologist whose investigation of a culture changes the culture.) For example, "If your last name has an odd number of letters, play a color other than the color of the top card; otherwise play the same color." It would be unfair, however, for a dealer not to tell the players when a rule of this tricky type is used.

The cards in the illustration have been played according to a simple rule not mentioned in this article. The reader may enjoy puzzling it out before it is explained. Note that the first seven cards follow a pattern of alternate colors. This often happens in a game as well as in the history of science. Players have in mind a condition that is not really part of the rule, but they stick by it until an experiment proves that the rule is simpler than they suspected or that their successes were merely accidental.

ADDENDUM

Although many games contain inductive features, only a few have sufficiently strong inductive aspects to justify calling them induction games. I can think only of Battleship (sometimes called Salvo), a children's pencil and paper game; Jotto and similar word games; and a parlor game called "Going on a Trip." This last game was called to my attention by I. Richard Lapidus of the physics department at Columbia University. The leader writes on a slip of paper a rule for determining what objects may be taken on a trip. He then says, "I plan to take a _____," naming an object that conforms to the rule. Guests take turns asking "Can I take a _____?" and are told by the leader whether the object they name is permitted. The first to guess the rule is the winner. Rules may be simple or complicated. A tricky rule: the object must begin with the same letter as the last name of the person taking it.

I suspect that there are many possibilities for unusual induction games that have not yet been explored – the guessing of concealed visual patterns, for example. Imagine a square-shaped box into which 100 square tiles will fit. Six hundred tiles are available, colored on one side, black on the other. There are six different colors, 100 tiles of each color. The leader secretly places 100 tiles in the box, forming a pattern that is strongly ordered (patterns can vary from one solid color to very complicated structures). He turns the box upside-down on the table, then removes it, leaving the tiles in square formation, black sides up. Players take turns choosing a single tile and reversing it. The first person to sketch a correct picture of the entire pattern is the winner. Players should sketch their guesses without letting other players see them, showing them only to the leader.

In playing Eleusis, the tendency to think of the dealer as God is so strong that players often find themselves drifting into a kind of theological lingo. A deal may be spoken of as a player's "turn to be God." If a dealer makes a mistake and violates his own rule by calling a card right that should have been wrong, the event is spoken of as a "miracle." Robert Abbott recalls one game in which the dealer, seeing that no one was capable of guessing his rule, pointed to a card in a player's hand and said, "Play that one."

"I've just had a divine revelation," the player responded.

ANSWERS

The secret rule determining the order of the cards in Figure 76 is: "Play a club or diamond if the top card of the pile is even; a heart or spade if the card is odd."

It is possible to formulate an infinity of other rules. Howard Givner of Brooklyn; Gerald Wasserman of Woodmere, New York; and Federico Fink of Buenos Aires suggested this one: "Play any card that differs in value from the top card of the pile." This is a simpler rule, but if correct it is difficult to explain how the stronger ordering of the cards, expressed by the first rule, could have come about. It is possible that all players erroneously guessed the first rule and played accordingly, and no one happened to play a card that matched in value the top card of the pile. In actual play, of course, the discarded cards provide additional clues for distinguishing between rival hypotheses.

C. A. Griscom, of New York, N. Y., was one of several readers who thought of extremely complicated rules. Griscom's rule concerns only the values of the cards, and assumes that the ace has a value of 14. No "going around the corner" is permitted. Play a card that either is larger or smaller than the top card of the pile, but if you continue the direction of change adopted by the previous player, you must increase the increment of change. If a larger increment is impossible, then the increment is given a value of 1.

It is an important insight into scientific method to realize that many hypotheses can be formulated to explain a given set of facts, and that any hypothesis can always be patched up, so to speak, to fit new facts that contradict it. For instance, if someone were to play the Eight of Diamonds on the Eight of Clubs, the last rule could be saved by adding that the Eight of Diamonds was an exceptional card that could be played at any time. Many a scientific hypothesis (e.g., the Ptolemaic model of the universe) has been elaborated to a fantastic degree in efforts to accommodate embarrassing new facts before it finally gave way to a simpler explanation.

All of which raises two profound questions in the philosophy of science: Why is the simplest hypothesis the best choice? How is "simplicity" defined?

POSTSCRIPT

Over the years Robert Abbott kept improving the rules of his game. His final version, which I called "the new Eleusis," was the topic of a later column reprinted in Book 13 (Chapter 16). Sidney Sackson's induction board game, Patterns, is described in Chapter 4 of Book 8. My *Discover* article, "The Computer as Scientist," is reprinted in *Gardner's Whys and Wherefores* (University of Chicago Press, 1984). It concerns computer algorithms capable of discovering laws of physics by surveying empirical data.

For centuries, philosophers of science have struggled with the problem, raised by David Hume, of justifying the success of induction. I side with those who agree with John Stuart Mill that the *only* way to justify induction is to assume that nature is patterned. Of course, this assumption is based on induction, but the circle is not vicious. It is unassailable. Bertrand Russell not only eventually came to this conclusion, but in his last major work, *Human Knowledge, Its Scope and Limits*, he tried to state a minimum number of posits that would describe how the universe is patterned.

BIBLIOGRAPHY

Delphi: A Game of Inductive Reasoning. Martin D. Kruskal. Plasma Physics Laboratory, Princeton University, 1962. A monograph of 16 pages.
The New Eleusis. Robert Abbott. Privately published, 1977.

"Eleusis: The Game with the Secret Rule." Sid Sackson in *Games* (May–June 1978): 18–19.

"Simulating Scientific Inquiry with the Card Game Eleusis." H. Charles Romesburg in *Science Education* 3 (1979): 599–608.

"The Methodology of Knowledge Layers for Inducing Descriptions of Sequentially Ordered Events." Thomas Dietterich. Department of Computer Science, University of Illinois at Urbana, master's thesis, 1980.

New Rules for Classic Games. R. Wayne Schmittberger. Wiley, 1992.

CHAPTER SIXTEEN

Origami

THE ORIGINS OF ORIGAMI are lost in the haze of early Asian history. Folded-paper birds appear as kimono decorations in eighteenth-century Japanese prints, but the art is certainly many centuries older in both China and Japan. At one time it was considered an accomplishment of refined Japanese ladies; now its chief practitioners seem to be the geisha girls and the Japanese children who learn it in school. During the past 20 years, there has been a marked upsurge of interest in origami in Spain and South America. The great Spanish poet and philosopher Miguel de Unamuno helped pave the way by writing a mock-serious treatise on the subject and by developing a basic fold that led to his invention of many remarkable new origami constructions.

Traditionally, origami is the art of folding realistic animals, birds, fish, and other objects from a single sheet of paper, without cutting, pasting, or decorating. In modern origami these restrictions are sometimes bypassed – a small scissor snip here, a dab of paste there, a penciled pair of eyes and so on. But just as the charm of Asian poetry lies in suggesting as much as possible with a minimum of words and within a rigid framework of rules, so the attraction of origami lies in the extraordinary realism that can be obtained with nothing more than a square of paper and a pair of deft hands. A sheet is folded along dull geometrical lines. Suddenly it is transformed into a delicate piece of miniature semiabstract sculpture that is often breathtakingly lovely.

In view of the geometrical aspect of paper folding, it is not surprising that many mathematicians have been fascinated by this whimsical, gentle art. Lewis Carroll, for example, who taught mathematics at Oxford, was an enthusiastic paper folder. (His diary records the occasion on which he first learned with delight how to fold a device that made a loud pop when it was swished through the air.) The literature of recreational mathematics includes many booklets and articles on folded-paper models, including those curious toys called flexagons.

The very act of folding raises an interesting mathematical question. Why is it that when we fold a sheet of paper the crease is a straight line? High school geometry texts sometimes cite this as an illustration of the fact that two planes intersect in a straight line, but this is clearly not correct because the parts of a folded sheet are parallel planes. Here is the proper explanation, as given by L. R. Chase in *The American Mathematical Monthly* for June–July 1940.

> Let p and p' be the two points of the paper that are brought into coincidence by the process of folding, then any point a of the crease is equidistant from p and p', since the lines ap and ap' are pressed into coincidence. Hence the crease, being the locus of such points a, is the perpendicular bisector of pp'.

The folding of regular polygons, though not a part of classic origami, is a challenging classroom exercise. The equilateral triangle, square, hexagon, and octagon are quite easy to fold, but the pentagon offers special difficulties. The simplest way to do it is to tie a knot in a strip of paper and press it flat (see illustration at left in Figure 77). This model conceals a topper. If we fold over one end of the strip and hold the knot up to a strong light (see illustration at right), we see the famous pentagram of medieval witchcraft.

Paper can also be folded to produce tangents that have as their envelope various low-order curves. The parabola is particularly easy to demonstrate. We first mark a point a few inches from one edge of the paper; then we crease the paper about 20 times at various spots, making sure that each crease is made when the edge is folded so that the edge intersects the point. Figure 78 shows the striking illusion of the parabola that results. The point is the focus of the curve, the edge of the paper is its directrix, and each crease is tangent to the curve. It is easy to see that this method of folding ensures that every point on the curve is equally distant from the focus and the directrix, a property that defines the parabola.

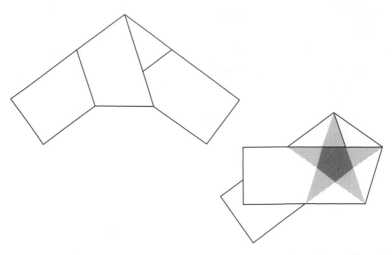

Figure 77. A strip is folded in a pentagon by tying it in a knot (left). If the strip is folded again, and held up to the light, a pentagram appears. (Artist: James Egleson)

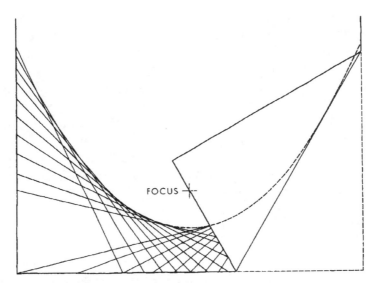

Figure 78. The tangents of a parabola are formed by folding the bottom edge of paper to the focus. (Artist: James Egleson)

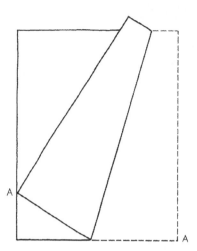

Figure 79. A calculus problem in paper folding. (Artist: James Egleson)

Closely related to this folding procedure is an interesting problem in elementary calculus. Suppose we have a sheet of paper that
is 8 × 11 inches in size. We fold it so that corner A (see Figure 79) just
touches the left edge. By moving the corner up and down the edge,
creasing at each position, we obtain tangents to a parabola that has
corner A for its focus. At what spot along the left edge must corner
A be placed so that a crease that intersects the bottom edge will be
as short as possible? What is the length of such a crease? Readers
unfamiliar with calculus may enjoy tackling the following simpler
variation. If the paper's width is reduced to 7.68 inches and the corner is folded to a spot 5.76 inches above the base, exactly how long
will the crease be?

And now, without apologies, I leave the more mathematical
aspects of paper folding to explain how to make what is in many
ways the most remarkable of all origami constructions: the bird that
flaps its wings. This object is both a thing of beauty and a mechanical masterpiece. The reader is urged to take a square of paper (patterned wrapping paper is excellent) and master the intricate folds.

A square eight inches on a side is a convenient size to use. (Some
experts like to make a miniature bird from a dollar bill that is first
folded into a square.) Crease the sheet along the two diagonals,
then turn it over (1 in Figure 80) so that the "valley folds" become

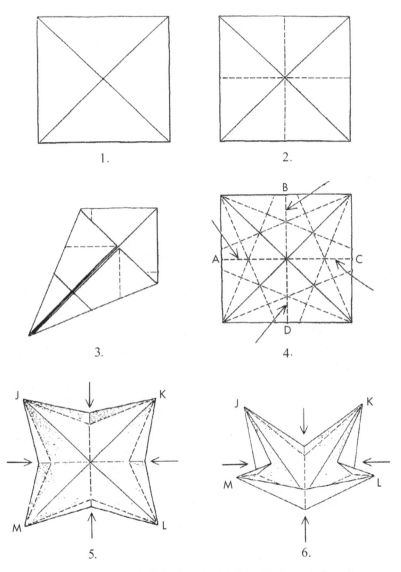

Figure 80. How to fold the flapping bird. (Artist: James Egleson)

"mountain folds." (In the illustrations all valley folds are shown as broken lines; all mountain folds as solid lines.)

Fold the paper in half, unfold, then fold in half the other way and unfold. This adds the two valley folds shown at 2 in the illustration.

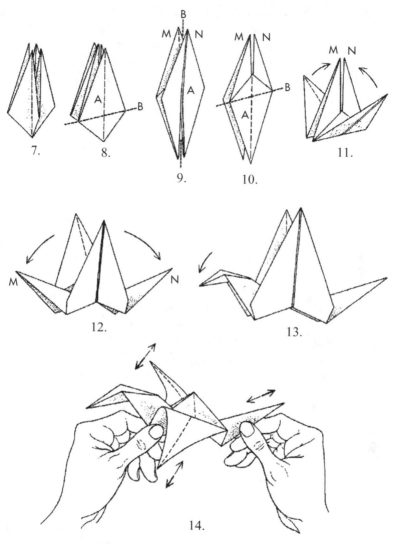

Figure 80. (Continued from page 164.)

Fold two adjacent sides over to meet (3 in illustration). Unfold, and then do the same thing at each of the other three corners. The paper will now be creased as shown at 4. (Note that the creases outline a regular octagon in the center of the square.)

The next step is extremely difficult to describe, though it is easily done once you get the hang of it. Note the four short valley-segments indicated by arrows at 4 in the illustration. Pinch these segments so that they become mountain folds. The centers of each side (labeled A, B, C, and D at 4) are pushed inward. The result is shown at 5. This raises the corners of the square (labeled J, K, L, and M) so that an oblique view of the model now appears as at 6.

If all the folds are in neat order (be sure the center of the square is pushed down as far as it will go), it should now be easy to bring all four corners together at the top as illustrated at 7. Flatten the model by bringing the sides together as shown at 8.

Flap A (at 8) is folded down along the line B. Turn the paper over and do the same on the other side. The paper now has the form shown at 9.

Flap A (at 9) is folded to the left along vertical line B. Turn the model over and do the same on the other side. The result is depicted at 10.

Flap A (at 10) is folded up along line B. Turn the model over and repeat on the other side. Hold the resulting isosceles triangle so that it points upward (11). For the remaining steps it will be more convenient to hold the model in the air rather than to rest it on a table.

Pull M to the angle shown at 12 and press the paper flat at the base. Do the same with N. Now push down the corner of M, reversing the fold, and press flat to form the bird's head (13).

Shape the wings (do not fold them) so that from their base to top they curve slightly outward and forward. Hold the bird as shown at 14. When you pull gently on the tail, the wings flap gracefully.

A number of origami animals have action features: a fish that opens its mouth, a frog that hops when its back is stroked, and so on. Unamuno's translator tells us that the Spanish writer liked to fold such animals while he sipped his midday coffee in a Salamanca café. Little wonder that wide-eyed street urchins kept their noses glued to the window panes!

ADDENDUM

New books on paper folding are being written every year, and several origami construction kits are now on sale in the United States.

Some kindergarten and primary grade teachers are beginning to discover the art, but perhaps most teachers are still allergic to it because they associate it with the sterile practice, so widespread in kindergartens early in the century, of folding elaborate designs from colored paper. (The practice had been introduced by Friedrich Froebel, German founder of the kindergarten, and many U.S. teachers came under its baleful influence.)

The flapping bird was first described in English in *Half Hours of Scientific Amusement,* by Gaston Tissandier, London, 1890 (a translation of an 1889 French book). There is a simpler way to fold the bird than the one I chose for this chapter, but it is more difficult to explain in print.

The description of Unamuno folding animals in a Spanish restaurant appears in the English translation of his *Essays and Soliloquies* (Knopf, 1925). Ortega y Gasset, in a book about his friend Unamuno, tells of the occasion on which the philosopher folded some paper animals for a small boy who asked, "Do the little birds speak?" The question inspired one of Unamuno's best-known poems. His humorous essay on paper folding is in *Amor y pedagogia* (Barcelona, 1902). A more important article by Unamuno on paper folding appears in the Argentine magazine *Caras y caretas,* March 1, 1902.

Akira Yoshizawa of Tokyo is considered the world's greatest living origami artist. He has written several books on the subject, and many articles for Japanese newspapers and magazines. In South America, the best origami manuals are by Vicente Solórzano Sagredo, a dentist in Buenos Aires. There is an extensive literature on the art in both Japanese and Spanish, but I have confined the references in the bibliography for this chapter to books in English that are not too difficult to find.

ANSWERS

The problem of the folded sheet is best handled as a maxima-minima problem in calculus. If x be the distance from corner A (the corner that is folded over) to where the crease strikes the bottom edge, then $8 - x$ will be the distance remaining on the bottom edge. The distance from the lower left corner to the point where corner A touches the left edge will be $4\sqrt{x - 4}$, the distance from the corner A to the spot where the crease strikes the right edge will be $2x/\sqrt{x-4}$, and the crease itself will be $\sqrt{x^3}/\sqrt{x-4}$. If the derivative of this last function is equated to zero, x will have a value of 6. The corner therefore touches the side edge at a point $4\sqrt{2}$ above the bottom, and the crease will be $6\sqrt{3}$ or a little more than 10.392 inches.

The interesting feature of this problem is that, regardless of the paper's width, the minimum crease intersecting the bottom edge is obtained by folding so that x is exactly three fourths of the paper's width. This three-quarter length multiplied by the square root of three gives the length of the crease. If the value to be minimized is the *area* of the part folded over, then x is always two thirds of the paper's width.

The crease in the simpler problem (in which the paper's width is 7.68 and the corner is folded to a point 5.76 above the base) is exactly 10 inches long.

POSTSCRIPT

Some 50 years ago I was asked to write a short account of paper folding for a new edition of the *Encyclopedia Britannica*. Several years went by before the set was published. By that time my piece was hopelessly out of date. An explosion of the nation's interest in origami had taken place, thanks in large part to a remarkable woman named Lillian Oppenheimer. She sponsored origami workshops, lectured on the topic, appeared on TV, even edited *The Origamian*, a periodical on origami.

On a trip to Japan, Mrs. Oppenheimer's TV appearances sparked a similar revival in Japan. The ancient art of paper folding had degenerated to where it was practiced only by geisha girls. Lillian located Japan's top origami expert, then living in poverty, and established a fund that allowed him to live comfortably.

I had the pleasure of knowing Lillian, and even contributed a paper fold to an origami exhibit she sponsored at Manhattan's Cooper Union Museum. It was a bird that balanced on a bottle top – a balance made possible by two pennies concealed inside the bird's wing tips. At the exhibit I was honored to meet the daughter of

Miguel de Unamuno, a Spanish philosopher who is one of my heroes. He was a skilled paper folder who created several new fundamental folds. Incidentally, Mrs. Oppenheimer had three sons by a previous marriage all of whom became distinguished mathematicians: William Kruskal at the University of Chicago, Joseph Kruskal at Bell Labs, and Martin Kruskal at Princeton University.

The flapping bird continues to be the most impressive, most beautiful of all origami action toys. I was amazed to learn that it was not invented in Japan, but in Europe. There must be earlier accounts of it in European literature than the one I found in Tissandier's book. Whoever created it surely deserves recognition.

There are now more than 30,000 origami figures described in books and articles, and mathematicians have developed an extensive mathematics of folded structures. In 2006 the fourth international conference on origami was held at the California Institute of Technology. This great upsurge of interest in paper folding has produced a raft of books on the topic, as well as kits with special paper squares. Paper differently colored on its two sides make possible figures that are strikingly bicolored, like black and white penguins.

Origami USA, 15 West 77th Street, New York, NY 10024, sponsors conventions, sells supplies, and publishes literature. It has a Web site.

BIBLIOGRAPHY

Geometrical Exercises in Paper Folding. T. Sundara Row. Madras, 1893. The fourth revised edition was reissued in 1958 by The Open Court Publishing Co., La Salle, Illinois.

"The Art of Paper Folding in Japan." Frederick Starr in *Japan* (October 1922).

Fun with Paper Folding. William D. Murray and Francis J. Rigney. Revell, 1928. Reprinted by Dover Publications, 1960, and retitled *Paper Folding for Beginners.*

Paper Toy Making. Margaret W. Campbell. Pitman, 1937.

Paper Magic: The Art of Paper Folding. Robert Harbin. Oldbourne Press, 1956.

Paper Folding for the Mathematics Class. Donovan A. Johnson. National Council of Teachers of Mathematics, 1957.

How to Make Origami. Isao Honda. McDowell, Obolensky, 1959.

Plane Geometry and Fancy Figures: An Exhibition of the Art and Technique of Paper Folding. Introduction by Edward Kallop. Cooper Union Museum, 1959.

Fun-Time Paper Folding. Elinor Massoglia. Children's Press, 1959.

Origami for the Connoisseur. Kunihiko Kasahara and Toshie Takahama. Japan Publications, 1987.

Folding the Universe. Peter Engel. Vintage, 1989.

Unit Origami. Tomoko Fusè. Japan Publications, 1990.

Origami Zoo. Robert Lang and Stephen Wise. St. Martin's, 1990.

Origami Animals. Hector Rojas. Sterling, 1993.

Origami Plain and Simple. Thomas Hull and Robert Neale. St. Martin's, 1994.

Secrets of Origami. Robert Harbin. Dover Publications, 1997.

Mathematical Origami. David Mitchell. Tarquin, 1997.

Russian Origami. Thomas Hull. St. Martin's, 1998.

Origami Omnibus. Kunihiko Kasahara. Japan Publications, 1998.

"A Mathematical Theory of Origami Constructions and Numbers." R. C. Alperin in *New York Journal of Mathematics* 6 (2000): 119–133.

Origami: The Complete Practical Guide to the Ancient Art of Paperfolding. Rick Beech. Lorenz Books, 2001.

Origami³. Thomas Hull. A K Peters, 2002.

Origami 1–2-3. David Petty. Sterling, 2002.

Origami Design Secrets. Robert Lang. A K Peters, 2003. This mammoth book of 585 pages is an awesome, comprehensive treatise on the mathematical aspects of origami. Its bibliography runs to fifteen pages. There is an eight-page glossary of terms, and a lengthy index.

"Origami Quiz." Thomas Hull in *Mathematical Intelligencer* 26: 4 (2004).

"Origami: Complexity in Creases (Again)." Robert Lang in *Engineering and Science* LXVII:1 (2004): 9–19. This marvelous article on origami mathematics contains beautiful full-color photos of Lang's incredible hummingbird and trumpet vine, his cicada and scorpion, and a dinosaur on the magazine's cover.

The Encyclopedia of Origami. Nick Robinson. Running Press, 2004.

"Cones, Curves, Shells, Towers: He Made Paper Jump to Life." Margaret Wertheim in *The New York Times*, June 22, 2004. This article, with color photographs, deals with the astonishing origami constructions of computer scientist David Huffman, former professor at the University of California, Santa Cruz.

"Origami as the Shape of Things to Come." Margaret Wertheim in *The New York Times*, February 15, 2005. This lengthy article is about the fantastic work of Dr. Erik Demaine, computer scientist at the

Massachusetts Institute of Technology, and a leading theoretician of origami mathematics. The field is turning out to have surprising applications to several fields of science, notably to the unsolved problem of how proteins fold so rapidly into well-defined structures.

"Folding Optical Polygons from Squares." David Dureisseix in *Mathematics Magazine* 79 (October 2006): 272–280.

Project Origami: Activities for Exploring Mathematics. Thomas Hull. A K Peters, 2006.

Origami A-B-C. David Petty. Sterling, 2006.

Marvelous Modular Origami. Meenakshi Mukerji. A K Peters, 2007.

"The Origami Lab." Susan Orlean in *The New Yorker* (February 19 and 26, 2007): 112–120. On physicist Robert Lang and his incredible origami constructions.

DOVER PAPERBACKS

Multimodular Origami Polyhedra. Rona Gurkewitz and Bennett Arnstein.

3-D Geometric Origami. Rona Gurkewitz and Bennett Arnstein.

Origami Insects. Robert Lang.

The Complete Book of Origami. Robert Lang.

Origami Inside-Out. John Montroll.

Teach Yourself Origami. John Montroll.

Bringing Origami to Life. John Montroll.

African Animals in Origami. John Montroll.

Animal Origami for the Enthusiast. John Montroll.

Animal Origami Adventure. John Montroll.

Dollar Bill Origami. John Montroll.

Prehistoric Origami. John Montroll.

A Plethora of Polyhedra in Origami. John Montroll.

A Constellation of Origami Polyhedra. John Montroll.

Origami Sculptures. John Montroll and Andrew Montroll.

Fascinating Origami. Vicente Palacios.

Origami From Around the World. Vincente Palacios.

Modular Origami Polyhedra. Lewis Simon, Bennett Arnstein, and Rona Gurkowitz.

The British Origami Society (BOS) is a charity organization that has published some hundred booklets on different aspects of origami. In 2006 they issued Robert Neale's *Which Came First?*, his latest book on action toys. Neale is best known in magic circles for his "bunny bill," a top hat folded with a dollar bill. When you squeeze the hat's side, the head of a rabbit pops up, seemingly out of the hat. Here is a list of Neale's books on action origami, a field in which he is the world's most creative inventor.

Bunny Bill. Magic, Inc., 1964.

Robert E. Neale's Trapdoor Card. Karl Fulves, 1983.

Origami, Plain and Simple. Robert Neale and Thomas Hull. St. Martin's Press, 1994.

Folding Money Fooling: How to Make Entertaining Novelties from Dollar Bills. Kaufman and Company, 1997.

Frog Tales: How to Fold Jumping Frogs from Poker Cards and Do Five Tricks with Them. H & R Magic Books, 2004.

Which Came First?: A Collection of Magical Designs by Bob Neale. British Origami Society, 2006.

Celebration of Sides: The Nonsense World of Robert Neale (DVD with Michael Weber). Murphy Magic Supplies, 2006.

CHAPTER SEVENTEEN

Squaring the Square

CAN A SQUARE *be subdivided into smaller squares of which no two are alike? This enormously difficult problem was long thought to be unsolvable, but now it has been defeated by translating it into electrical-network theory, then back into plane geometry again. Here William T. Tutte, associate professor of mathematics at the University of Toronto, presents a fascinating account of how he and three fellow students at the University of Cambridge finally squared the square.*

This is the story of a mathematical research conducted by four students of Trinity College, Cambridge, in the years 1936–1938. One was the author of this article. Another was C. A. B. Smith, now a statistical geneticist at the University of London. He is also well-known as a writer on the theory of games and the counterfeit-coin problem. Another was A. H. Stone, now researching at Manchester into recondite regions of point-set topology. He is one of the inventors of the flexagons described in [Gardner's Book 1]. The fourth was R. L. Brooks. He has now left the academic world for the Civil Service. But he retains an enthusiasm for mathematical recreations, and an important theorem in the theory of graph colorings bears his name. These four students referred to themselves, with characteristic modesty, as the "Important Members" of the Trinity Mathematical Society.

In 1936 there were a few references in the literature to the problem of cutting up a rectangle into unequal squares. Thus it was known that a rectangle of sides 32 and 33 units can be dissected into nine squares with sides of 1, 4, 7, 8, 9, 10, 14, 15, and 18 units (Figure 81). Stone was intrigued by a statement in Dudeney's *Canterbury Puzzles* which seemed to imply that it is impossible to cut up a *square* into unequal smaller squares. He tried to prove

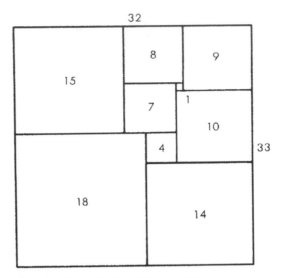

Figure 81. (Artist: James Egleson)

the impossibility for himself, but without success. He did, however, discover a dissection of the rectangle of sides 176 and 177 into 11 unequal squares (Figure 83).

This partial success fired the imaginations of Stone and his three friends and soon they were spending much time constructing, and arguing about, dissections of rectangles into squares. Any rectangle cut up into unequal squares was called by them a "perfect" rectangle. Years later the term "squared rectangle" was introduced to describe any rectangle cut up into two or more squares, not necessarily unequal.

The construction of perfect rectangles proved to be quite easy. The method used was as follows. First we sketch a rectangle cut up into rectangles, as in Figure 82. We then think of the diagram as a bad drawing of a squared rectangle, the small rectangles being really squares, and we work out by elementary algebra what the relative sizes of the squares must be on this assumption. Thus in Figure 82, we have denoted the sides of two adjacent small squares by x and y. We can then say that the side of the square immediately below them is $x + y$ and then that the side of the square next on the left is $x + 2y$, and so on. Proceeding in this way, we get the formulas shown

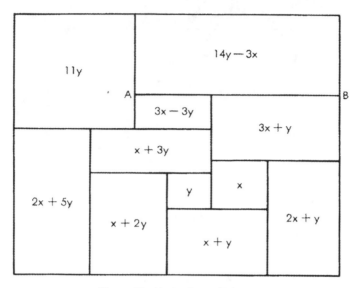

Figure 82. (Artist: James Egleson)

in Figure 82 for the sides of the 11 small squares. These formulas make the squares fit together exactly except along the one segment AB. But we can make them fit on AB too by choosing x and y to satisfy the equation $(3x + y) + (3x - 3y) = (14y - 3x)$, that is, $16y = 9x$. Accordingly we put $x = 16$ and $y = 9$. This gives the perfect rectangle of Figure 83, which is the one first found by Stone.

Sometimes this method gave negative values for the sides of some small squares. It was found, however, that such negative squares could always be converted into positive ones by minor modifications of the original diagram. They therefore gave no trouble. In some of the more complicated diagrams, it proved necessary to start with three unknown squares, with sides x, y, and z, and solve two linear equations instead of one at the end of the algebraic computations. Sometimes the squared rectangle finally obtained proved not to be perfect, and the attempt was considered a failure. Fortunately this did not happen very often. We recorded only "simple" perfect rectangles, that is, perfect rectangles containing no smaller ones. For example, the perfect rectangle obtained from Figure 81 by erecting a new component square of side 32 on the upper horizontal side is not simple, and we did not include it in our catalog.

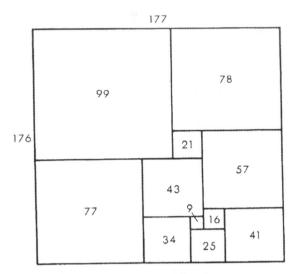

Figure 83. (Artist: James Egleson)

In this first stage of the research, large numbers of perfect rectangles were constructed in which the number of component squares ranged from 9 to 26. In the final form of each rectangle, the sides of the component squares were represented as integers without a common factor. Of course we all hoped that if we constructed enough perfect rectangles by this method we would eventually obtain one that was a "perfect square." But as the list of perfect rectangles lengthened, this hope faded. Production slowed down accordingly.

Inspection of the catalog we had constructed revealed some very odd phenomena. We had classified our rectangles according to their "order," that is, the number of component squares. We noticed a tendency for numbers representing sides to be repeated in any one order. Moreover the semiperimeter of a rectangle in one order often reappeared several times as a side in the next order. For example, using the full information now available, one finds that four of the six simple perfect rectangles of order 10 have semiperimeter 209 and that five of the 22 simple perfect rectangles of order 11 have 209 as a side. There was much discussion of this "Law of Unaccountable Recurrence," but it led to no satisfactory explanation.

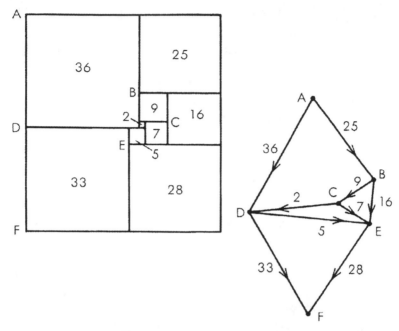

Figure 84. (Artist: James Egleson)

In the next stage of the research, we abandoned experiment in favor of theory. We tried to represent squared rectangles by diagrams of different kinds. The last of these diagrams, introduced by Smith, was a really big step forward. The other three researchers called it the Smith diagram. But Smith objected to this name, alleging that his diagram was only a minor modification of one of the earlier ones. However that may be, Smith's diagram suddenly made our problem part of the theory of electrical networks.

Figure 84 shows a perfect rectangle together with its Smith diagram. Each horizontal line segment in the drawing of the rectangle is represented in the Smith diagram by a dot, or "terminal." In the Smith diagram, the terminal is made to lie on a continuation to the right of its corresponding horizontal segment in the rectangle. Any component square of the rectangle is bounded above and below by two of the horizontal segments. Accordingly, it is represented by a line, or "wire," in the diagram joining the two corresponding terminals. We imagine an electric current flowing in each wire. The magnitude of the current is numerically equal to the side of the corresponding square, and its direction is from the terminal representing the upper horizontal segment to the terminal representing the lower one.

The terminals corresponding to the upper and lower horizontal sides of the rectangle may conveniently be called the positive and negative poles, respectively, of the electrical network.

Surprisingly enough the electric currents assigned by the above rule really do obey Kirchhoff's laws for the flow of current in a network, provided that we take each wire to be of unit resistance. Kirchhoff's first law states that, except at a pole, the algebraic sum of the currents flowing to any terminal is zero. This corresponds to the fact that the sum of the sides of the squares bounded below by a given horizontal segment is equal to the sum of the sides of the squares bounded above by the same segment, provided of course that the segment is not one of the horizontal sides of the rectangle. The second law says that the algebraic sum of the currents in any circuit is zero. This is equivalent to saying that when we describe the circuit, the net corresponding change of level in the rectangle must be zero.

The total current entering the network at the positive pole, or leaving it at the negative pole, is evidently equal to the horizontal side of the rectangle, and the potential difference between the two poles is equal to the vertical side.

The discovery of this electrical analogy was important to us because it linked our problem with an established theory. We could now borrow from the theory of electrical networks and obtain formulas for the currents in a general Smith diagram and the sizes of the corresponding component squares. The main results of this borrowing can be summarized as follows. With each electrical network, there is associated a number calculated from the structure of the network, without any reference to which particular pair of terminals is chosen as poles. We called this number the complexity of the network. If the units of measurement for the corresponding rectangle are chosen so that the horizontal side is equal to the complexity, then the sides of the component squares are all integers. Moreover, the vertical side is equal to the complexity of another network obtained from the first by identifying the two poles.

The numbers giving the side of the rectangle and its component squares in this system of measurement were called the "full" sides and "full" elements of the rectangle, respectively. For some rectangles the full elements have a common factor greater than unity. In any case, division by their common factor gives the "reduced" sides and elements. It was the reduced sides and elements that had been recorded in our catalog.

These results imply that if two squared rectangles correspond to networks of the same structure, differing only in the choice of poles, then the full horizontal sides are equal. Further, if two rectangles have networks that acquire the same structure when the two poles of each are identified, then the two vertical sides are equal. These two facts explained all the cases of "unaccountable recurrence" that we had encountered.

The discovery of the Smith diagram simplified the procedure for producing and classifying simple squared rectangles. It was an easy matter to list all the permissible electrical networks of up to 11 wires, and to calculate all the corresponding squared rectangles. We then found that there were no perfect rectangles below the ninth order, and only two of the ninth (Figures 81 and 84). There were six of the tenth order and 22 of the eleventh. The catalog then advanced, though more slowly, through the twelfth order (67 simple perfect rectangles) and into the thirteenth.

It was a pleasing recreation to work out perfect rectangles corresponding to networks with a high degree of symmetry. We considered, for example, the network defined by a cube, with corners for terminals and edges for wires. This failed to give any perfect rectangles. However, when complicated by a diagonal wire across one face, and flattened into a plane, it gave the Smith diagram of Figure 85 and the corresponding perfect rectangle of Figure 86. This rectangle was especially interesting because its reduced elements are unusually small for the thirteenth order. The common factor of the full elements is 6. Brooks was so pleased with this rectangle that he made a jigsaw puzzle of it, each of the pieces being one of the component squares.

It was at this stage that Brooks's mother made the key discovery of the whole research. She tackled Brooks's puzzle and eventually succeeded in putting the pieces together to form a rectangle. But it was

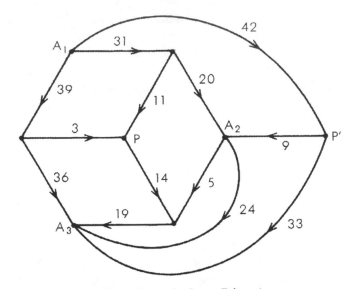

Figure 85. (Artist: James Egleson)

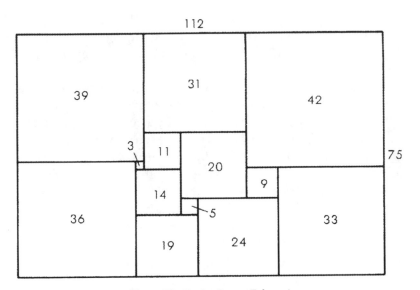

Figure 86. (Artist: James Egleson)

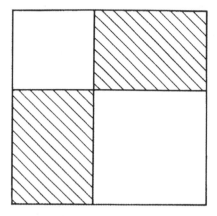

Figure 87. (Artist: James Egleson)

not the squared rectangle that Brooks had cut up! Brooks returned to Cambridge to report the existence of two different perfect rectangles with the same reduced sides and the same reduced elements. Here was unaccountable recurrence with a vengeance! The Important Members met in emergency session.

We had sometimes wondered whether it was possible for different perfect rectangles to have the same shape. We would have liked to obtain two such rectangles with no common reduced element, and thus get a perfect square by the construction shown in Figure 87. The shaded regions in this diagram represent the two perfect rectangles. Two unequal squares are then added to make the large perfect square. But no rectangles of the same shape had hitherto appeared in our catalog, and we had reluctantly come to believe that the phenomenon was impossible. Mrs. Brooks's discovery renewed our hopes, even though her rectangles failed in the worst possible way to have no common reduced element.

There was much excited discussion at the emergency session. Eventually the Important Members calmed down sufficiently to draw the Smith diagrams of the two rectangles. Inspection of these soon made clear the relationship between them.

The second rectangle is shown in Figure 88 and its Smith diagram in Figure 89. It is evident that the network of Figure 89 can be obtained from that of Figure 85 by identifying the terminals p

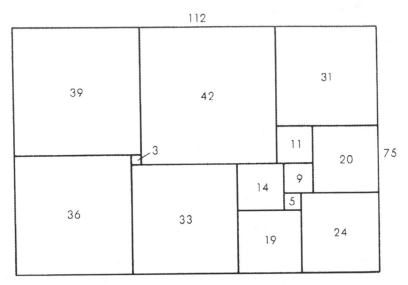

Figure 88. (Artist: James Egleson)

and p'. As p and p' happen to have the same electrical potential in Figure 85, this operation causes no change in the currents in the individual wires, no change in the total current, and no change in the potential difference between the poles. We thus have a simple

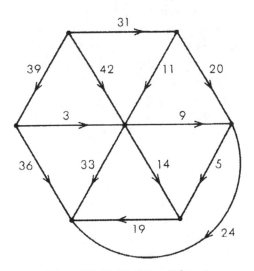

Figure 89. (Artist: James Egleson)

electrical explanation of the fact that the two rectangles have the same reduced sides and the same reduced elements.

But why do p and p' have the same potential in Figure 85? Before the emergency session broke up we had obtained an answer to this question also. The explanation depends on the fact that the network can be decomposed into three parts meeting only at the poles A_1 and A_2 and the terminal A_3. One of these parts consists solely of the wire joining A_2 and A_3. A second part is made up of the three wires meeting at p', and a third is constituted by the remaining nine wires. Now the third part has threefold rotational symmetry with p as the center of rotation. Moreover, current enters or leaves this part of the network only at A_1, A_2, and A_3, which are equivalent under the symmetry. This is enough to ensure that if any potentials whatever are applied to A_1, A_2, and A_3, the potential of p will be their average. The same argument applied to the second part of the network shows that the potential of p' must also be the average of the potentials of A_1, A_2, and A_3. Hence p and p' have the same potential, whatever potentials are applied to A_1, A_2, and A_3, and in particular they have the same potential when A_1 and A_2 are taken as poles in the complete network, and the potential of A_3 is fixed by Kirchhoff's laws.

The next advance was made accidentally by the present writer. We had just seen Mrs. Brooks's discovery completely explained in terms of a simple property of symmetrical networks. It seemed to me that it should be possible to use this property to construct other examples of pairs of perfect rectangles with the same reduced elements. I could not have explained how this would help us in our main object of constructing, or proving the impossibility of, a perfect square. But I thought we should explore the possibilities of the new ideas before abandoning them.

The obvious thing to do was to replace the third part of the network of Figure 85 by another network having threefold rotational symmetry about a central terminal. But this can be done only under severe limitations, which should now be explained.

It can be shown that the Smith diagram of a squared rectangle is always planar, that is, it can be drawn in the plane with no crossing wires. And the drawing can always be made so that no circuit separates the two poles. There is also a converse theorem that states that if an electrical network of unit resistances can be drawn in the plane

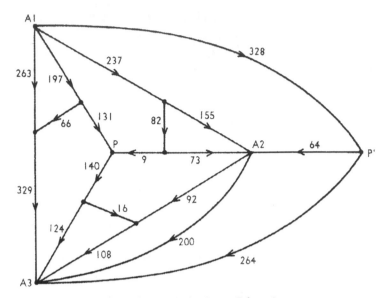

Figure 90. (Artist: James Egleson)

in this way, then it is the Smith diagram of some squared rectangle. It would not be proper to take up space in this book with rigorous proofs of these theorems. It would not even be historically accurate; the four researchers did without rigorous proofs right up to the time when they began to prepare their technical paper for publication.

It is not always advisable to disregard rigor in this way in the course of a mathematical research. In a research aiming at a proof of the Four Color Theorem, for example, such an attitude would be, and indeed often is, disastrous. But our research was largely experimental, and its experimental results were perfect rectangles. Our methods were justified, for the time being, by the rectangles they produced, even when their theory had not been precisely worked out.

But let us return to Figure 85 and the replacement of its third part by a new symmetrical network with center p. The complete network obtained in this way must not only be planar but it must remain planar when p and p' are identified.

After a few trials I found two closely related networks satisfying these conditions. The corresponding Smith diagrams are shown in Figures 90 and 91. As was expected, each diagram allowed of the

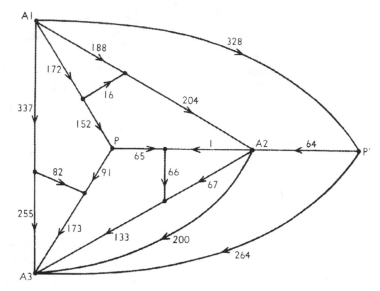

Figure 91. (Artist: James Egleson)

identification of *p* and *p'*, and so gave rise to two squared rectangles with the same reduced elements. But all four rectangles had the same reduced sides, and this result was quite unexpected.

Essentially the new discovery was that the rectangles corresponding to Figures 90 and 91 have the same shape, through they do not have their reduced elements all the same. A simple theoretical explanation of this was soon found. The two networks have the same structure, apart from the choice of poles, and therefore the rectangles have the same full horizontal side. Moreover the networks remain identical when poles are coalesced, and therefore the two rectangles have the same vertical side. We felt, however, that this explanation did not probe sufficiently deep, since it made no reference to rotational symmetry.

We eventually agreed to refer to the new phenomenon as "rotor-stator" equivalence. It was always associated with a network that could be decomposed into two parts, the "rotor" and the "stator," with the following properties. The rotor had rotational symmetry, the terminals common to the rotor and stator were all equivalent under the symmetry of the rotor, and the poles were terminals of the stator. In Figure 90, for example, the stator is made up of the three wires joining p' to A_1, A_2, and A_3, and the wire link-ing A_2 with A_3. A second network could then be obtained by an operation called "reversing" the rotor. With a properly drawn fig-ure this could be explained as a reflection of the rotor in a straight line passing through its center. Thus, starting with Figure 90, we can reflect the rotor in the line pA_3 and so obtain the network of Figure 91.

After studying a few examples of rotor-stator equivalence, the researchers convinced themselves that reversing the rotor made no difference to the full sides of the rectangle, and no difference to the currents in the wires of the stator. But the currents in the rotor might change. Satisfactory proofs of these results were obtained only at a much later stage.

Rotor-stator equivalence proved to have no very close relation-ship with the phenomenon discovered by Mrs. Brooks. It was merely another one associated with networks having a part with rotational symmetry. The importance to us of Mrs. Brooks's discovery was that it led us to study such networks.

A very tantalizing question now arose. What was the least pos-sible number of common elements in a rotor-stator pair of per-fect rectangles? Those of Figures 90 and 91 had seven common ele-ments, three of which corresponded to currents in the rotor. The same rotor with a stator consisting of a single wire A_2A_3 gave two perfect rectangles of the sixteenth order with four common ele-ments. Using a one-wire stator there seemed no theoretical rea-son why we should not obtain a pair of perfect rectangles having only one element, corresponding to the stator, in common. But we saw that if we could do this, we could also obtain a perfect square. For with the rotors of threefold symmetry that we were studying, a one-wire stator always represented a corner element of each corre-sponding rectangle. From two perfect rectangles with only a corner element in common, we can expect to obtain a perfect square by the construction illustrated in Figure 92. Here the shaded regions rep-resent the two rectangles. The square in which they overlap is the common corner element.

Naturally we got to work calculating rotor-stator pairs. We made the rotors as simple as we could, partly to save labor and partly in the hope of getting a perfect square with small reduced elements. But one construction after another failed because of common

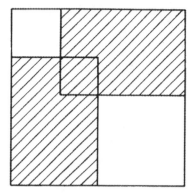

Figure 92. (Artist: James Egleson)

elements in the rotor, and we became discouraged. Was there some theoretical barrier still to be explored?

It occurred to some of us that perhaps our rotors were too simple. Something more complicated might be better. The numbers involved would be much bigger and the likelihood of a chance coincidence would be reduced. So it came to pass that Smith and Stone sat down to compute a complicated rotor-stator pair while Brooks, unknown to them, worked on another in a different part of the College. After some hours Smith and Stone burst into Brooks's room crying "We have a perfect square!" To which Brooks replied "So have I!"

Both these squares were of the sixty-ninth order. But Brooks went on to experiment with simpler rotors and obtained a perfect square of the thirty-ninth order. This corresponds to the rotor shown in Figure 93. A brief description of it is provided by the following formula:

[2378, 1163, 1098], [65, 1033], [737, 491], [249, 242], [7, 235],

[478, 259], [256], [324, 944], [219, 296], [1030, 829, 519, 697],

[620], [341, 178], [163, 712, 1564], [201, 440, 157, 31], [126, 409],

[283], [1231], [992, 140], [852].

In this formula each pair of brackets represents one of the horizontal segments in the subdivision pattern of the perfect square. These segments are taken in vertical order, beginning with the upper horizontal side of the square, and the lower horizontal side is omitted. The numbers enclosed by a pair of brackets are the sides

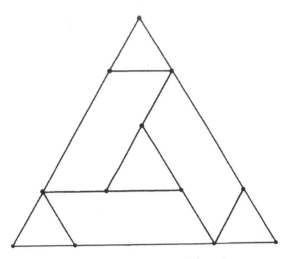

Figure 93. (Artist: James Egleson)

of those component squares that have their upper horizontal sides in the corresponding segment. They are taken in order from left to right. The reduced side of the perfect square is the sum of the numbers in the first pair of brackets, that is, 4639.

The notation we have just used is that of C. J. Bouwkamp. He has employed it in his published list of the simple squared rectangles up to the thirteenth order.

This really completes the story of how this particular team solved the problem of the perfect square. We did more work on the problem, it is true. All the perfect squares obtained by the rotor-stator method had certain properties that we regarded as blemishes. Each contained a smaller perfect rectangle; that is, was not simple. Each had a point at which four of the component squares met; that is, was "crossed." Finally, each had a component square, not one of the four corner squares, that was bisected by a diagonal of the complete figure. Using a more advanced theory of rotors we were able to get perfect squares without the first two blemishes. Years later, by a method based on a completely different kind of symmetry, I obtained a perfect square of the sixty-ninth order free of all three kinds of blemish. But for an account of this work, I must refer the interested reader to our technical papers.

There are three more episodes in the history of the perfect square that ought to be mentioned, though each one may seem like an anti-climax. To begin with, we kept adding to the list of simple perfect rectangles of the thirteenth order. Then one day we found that two of these rectangles had the same shape and no common element. They gave rise to a perfect square of the twenty-eighth order by the construction of Figure 87. Later we found a thirteenth-order perfect rectangle that could be combined with one of the twelfth order and one extra component square to give a perfect square of the twenty-sixth order. If the merit of a perfect square is measured by the small-ness of its order, then the empirical method of cataloging the perfect rectangles had proved superior to our beautiful theoretical method.

Other researchers have used the empirical method with spectac-ular results. R. Sprague of Berlin fitted a number of perfect rectan-gles together in a most ingenious way to produce a perfect square of the fifty-fifth order. This was the first perfect square to be published (1939). More recently T. H. Willcocks of Bristol, who did not confine his catalog to simple and perfect squared rectangles, obtained a per-fect square of the twenty-fourth order (Figure 94). Its formula is as follows: [55, 39, 81], [16, 9, 14], [4, 5], [3, 1], [20], [56, 18], [38], [30, 51], [64, 31, 29], [8, 43], [2, 35], [33]. This perfect square still holds the low-order record.

Unlike the theoretical method, the empirical one has not yet given rise to any simple perfect square.

In case any reader should like to do some work on perfect rectan-gles himself, here are two unsolved problems. The first is to deter-mine the smallest possible order for a perfect square. The second is to find a simple perfect rectangle whose horizontal side is twice the vertical side.

– W. T. Tutte

ADDENDUM

In 1960, C. J. Bouwkamp published a catalog of all simple squared rectangles (i.e., squared rectangles that do not contain smaller squared rectangles) through order 15. With the aid of an IBM-650

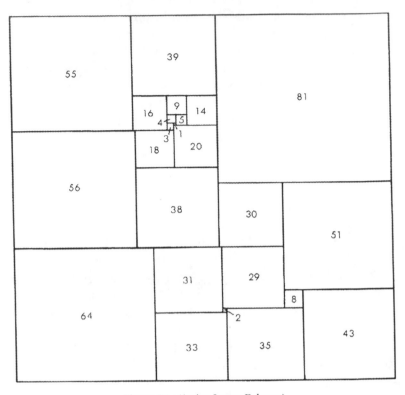

Figure 94. (Artist: James Egleson)

computer, Bouwkamp and his associates tabulated the following results:

Order of rectangle	9	10	11	12	13	14	15
Imperfect	1	0	0	9	34	104	283
Perfect	2	6	22	67	213	744	2,609

The imperfect simple squared rectangles are those containing at least two squares of the same size. The perfect ones are those in which the squares are all of different sizes. The total number of simple squared rectangles through order 15 is 4,094. It is interesting to note that no simple squared rectangles of orders 10 and 11 are possible without being perfect. The single imperfect simple rectangle of order 9 has the formula: [6, 4, 5], [3, 1], [6], [5, 1], [4]. It has a pleasing symmetry and makes an excellent dissection puzzle for a child.

Several squared rectangles appear in the puzzle books of Sam Loyd and H. E. Dudeney, but none that is either simple or perfect. A twenty-sixth-order squared square, perfect but not simple, is depicted in Hugo Steinhaus's *Mathematical Snapshots*, and Maurice Kraitchik's *Mathematical Recreations*. So far as I know, no squared rectangles have been marketed as dissection puzzles. One reader, William C. Spindler of Arlington, California, sent me a photograph of a handsome rectangular patio that he built with 19 square blocks of concrete separated by two-inch redwood strips.

The smallest published square that is both simple and perfect is a thirty-eighth-order square with a side of 4,920, discovered by R. L. Brooks. In 1959, this was bettered by T. H. Willcocks of Bristol, England, with a thirty-seventh-order square, 1,947 on the side. Is it possible to dissect a cube into a finite number of smaller cubes, all different sizes? No, and a beautiful proof of this is given by the "Important Members" in the second entry in the list of references. The proof runs as follows:

Imagine that you have before you, resting on a table, a cube cut into smaller cubes, no two the same size. The bottom face of this cube will of course be a squared square. Within this square will be a smallest square. It is easy to see that this smallest square cannot be touching an edge of the large square that is the cube's bottom face. Therefore the smallest cube that rests directly on the table top – we will call it cube A – must be surrounded by other cubes. None of the surrounding cubes can be smaller than cube A, therefore it will be surrounded by walls that rise above it. On cube A still smaller cubes will rest. They form a squared square on the top face of cube A. Within this squared square will be a smallest square, calling for a cube B that is the smallest cube resting directly on top of cube A.

The same argument in turn will call for a cube C that is the smallest cube resting on cube B. Thus we are faced with an endless regress of smaller and smaller cubes, like the fleas in Dean Swift's familiar jingle that have lesser fleas to bite 'em, and so on ad infinitum. No cube, therefore, can be dissected into a finite number of smaller cubes of different sizes.

A hypercube of four dimensions has "faces" that are cubes. If a hypercube could be hypercubed, then its faces would be cubed cubes; this is impossible, so it follows that no hypercube can be hypercubed. For similar reasons, no fifth-dimensional cubes can be cut into smaller fifth-dimensional cubes of different sizes, and so on for all cubes of higher dimensions.

For an example of a perfect squared rectangle of order infinity, see Figure 47 in Chapter 8.

POSTSCRIPT

The most significant new result on squared squares is finding the smallest order for a simple perfect squared square. It is 21. You'll find it in *The Journal of Combinatorial Theory*, Vol. 35B (1978), pp. 260–263 and Chapter 11 of my Book 14.

The first solution to the problem of finding a simple perfect rectangle with sides in a 2:1 ratio was published by R. L. Brooks in *The Journal of Combinatorial Theory*, Vol. 8 (1970), pp. 232–243. It has 1,323 squares! Examples of orders 23, 24, and 25 are given by P. J. Federico in the same issue. Federico's excellent history of the topic, "Squaring Rectangles and Squares," can be found in *Graph Theory and Related Topics*, edited by J. A. Bundy and V. K. Murty (Academic Press, 1979). Its bibliography lists 73 references!

Clifford Pickover, in *The Möbius Strip* (Avalon, 2006), pp. 105–106, reveals that five distinct squares will tile a rectangle that is a Möbius surface, but that nine tiles are needed for a cylindrical rectangle. He adds that he knows of no similar results for the Klein bottle or projective plane.

BIBLIOGRAPHY

"Beispiel Einer Zerlegung des Quadrats in Lauter Verschiedene Quadrate." R. Sprague in *Mathematische Zeitschrift* 45 (1939): 607–608.
"The Dissection of Rectangles into Squares." R. L. Brooks, C. A. B. Smith, A. H. Stone, and W. T. Tutte in *Duke Mathematical Journal* 7 (1940): 312–340.
"Question E401 and Solution." A. H. Stone in *American Mathematical Monthly* 47 (1940): 570–572.

"On the Dissection of Rectangles into Squares (I–III)." C. J. Bouwkamp in *Koninklijke Nederlandsche Akademie van Wetenschappen, Proceedings* 49 (1946): 1,176–1,188 and 50 (1947): 58–78.

"On the Construction of Simple Perfect Squared Squares." C. J. Bouwkamp in *Koninklijke Nederlandsche Akademie van Wetenschappen, Proceedings* 50 (1947): 1,296–1,299.

"A Simple Perfect Square." R. L. Brooks, C. A. B. Smith, A. H. Stone and W. T. Tutte in *Koninklijke Nederlandsche Akademie van Wetenschappen, Proceedings* 50 (1947): 1,300–1,301.

"The Dissection of Equilateral Triangles into Equilateral Triangles." W. T. Tutte in the *Proceedings of the Cambridge Philosophical Society* 44 (1948): 464–482.

"A Class of Self-Dual Maps." C. A. B. Smith and W. T. Tutte in *Canadian Journal of Mathematics* 2 (1950): 179–196.

"Squaring the Square." W. T. Tutte in *Canadian Journal of Mathematics* 2 (1950): 197–209.

"A Note on Some Perfect Squared Squares." T. H. Willcocks in *Canadian Journal of Mathematics* 3 (1951): 304–308.

Catalog of Simple Squared Rectangles of Orders Nine Through Fourteen and Their Elements. C. J. Bouwkamp, A. J. W. Duijvestijn, and P. Medema. Department of Mathematics, Technische Hogeschool, Eindhoven, Netherlands, 1960.

CHAPTER EIGHTEEN

Mechanical Puzzles

MECHANICAL PUZZLES, in contrast to the pencil-and-paper variety, are puzzles requiring some sort of special equipment that must be operated by hand. The equipment may be nothing more than a few pieces of cardboard, or it may be an elaborate construction of wood or metal that is beyond the ability of most home craftsmen to duplicate. Manufactured puzzles of the mechanical type, sold in toy and novelty shops, are often extremely interesting from a mathematical standpoint, and for this reason are sometimes collected by students of recreational mathematics. The largest such collection known to me is owned by Lester A. Grimes, a retired fire-protection engineer who lives in New Rochelle, New York (Figure 95). (A smaller collection, though stronger on nineteenth-century items and old Chinese puzzles, is owned by Thomas Ransom of Belleville, Ontario.) Grimes's collection numbers about 2,000 different puzzles, many of them exceedingly rare. The following account is based largely on this collection.

The tangram puzzle game originated in China in the early nineteenth century where it was called the *ch'i ch'iao* (the ingenious seven-piece plan). It quickly became a fad, not only in the Eastern countries but in Western nations as well. Napoleon is said to have whiled away his exiled hours with a set, now in a Paris museum. The name tangrams may have been coined by an anonymous toy manufacturer here or in England. Many books on tangram figures have been published, one a booklet by Sam Loyd who fabricated a history of the puzzle, falsely claiming it to be thousands of years old.

Dissection puzzle-games similar to tangrams have appeared from time to time (the ancient Greeks and Romans amused themselves with a 14-piece dissection of a rectangle, attributed to

Figure 95. Lester A. Grimes of New Rochelle, N.Y., and some of his 2,000 mechanical puzzles. (Photograph courtesy of Jerry Slocum.)

Archimedes), but tangrams has outlived them all. To understand why, you need only cut a set of "tans" from a square of heavy cardboard and then try your skill at solving a few tangram puzzles or devising some new ones. Figure 96 shows how the square is dissected. The rhomboid should be colored black on both sides so that it can be turned over if desired. All seven tans must be used in every figure. Only the geometrical patterns require a bit of effort to solve; a variety of picture-figures are included to show the graceful effects that can be achieved.

Simple dissection-puzzles of this type occasionally provoke mathematical problems that are far from trivial. Suppose, for example, you wish to find all the different convex polygons (polygons with no outside angles less than 180 degrees) that can be formed with the seven tans. You might find them by prolonged trial and error, but how can you prove that you have indeed discovered all of them? Two mathematicians at the National University of Chekiang – Fu Traing Wang and Chuan-Chih Hsiung – published a paper in 1942 on just

Figure 96. Chinese tangrams (top left) and some of the figures that can be made with the seven tans.

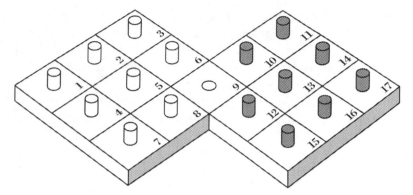

Figure 97. How can the black and white pegs be transposed in the smallest number of moves? (Artist: Harold Jacobs)

this problem. Their approach was ingenious. Each of the five largest tans can be divided into isosceles right-angle triangles congruent with the two small tans, so that altogether the seven tans are made up of 16 identical isosceles right-angle triangles. By a clever chain of arguments the two Chinese authors show that 20 different convex polygons (not counting rotations and reflections) can be formed with 16 such triangles. It is then easy to prove that exactly 13 of these 20 polygons are tangrams.

Of the 13 possible convex tangrams, one is a triangle, six are quadrilaterals, two are five-sided and four are six-sided. The triangle and three quadrilaterals are shown in Figure 96. It is a pleasant but by no means easy task to discover the other nine. Each can be formed in more than one way, but there is one hexagon that is considerably more elusive than the other 12 figures.

Another popular genus of mechanical puzzle, species of which can be traced back many centuries, involves counters or pegs that are moved across a board according to prescribed rules in order to achieve a certain result. One of the best puzzles of this type, widely sold in Victorian England, is shown in Figure 97. The object of the puzzle is to exchange the positions of the black and white pegs in the fewest number of moves. A move is either (1) from one square to an adjacent vacant square, or (2) a jump over an adjacent peg to a vacant square. A peg may jump a peg of the same or opposite color. All moves are "rook-wise"; no diagonal moves are permitted.

Most puzzle books give a solution in 52 moves, but Henry Ernest Dudeney, the English puzzle expert, discovered an elegant solution in 46 moves. The puzzle can be worked by placing small counters on top of the pegs in the illustration. The squares are numbered to facilitate recording the answer.

This and the preceding puzzle were singled out because the reader can construct them with little effort. Most of the puzzles in Grimes's collection cannot be made easily; since they must be handled to be appreciated, I shall content myself with a brief description of their variety. There are puzzles boxes, purses, and other containers to be opened by cleverly hidden methods, hundreds of odd-shaped wire puzzles to be taken apart, silver bracelets and finger rings made of separate pieces that interlock ingeniously, cords to be removed from objects without cutting or untying, glass-topped dexterity puzzles containing objects that are rolled or shaken into desired positions, rings to be removed from rods, eggs to be balanced on end, mazes in three dimensions, Chinese puzzles of interlocked wooden pieces, items involving moving counters and sliding blocks, and hundreds of curious puzzles that defy all classification. Who invents such toys? To trace them back to their origins would be an impossible task. In most cases, it is not even known in what country a puzzle originated.

There is one happy exception. A section of Grimes's collection is reserved for about 200 remarkable puzzles invented and constructed by L. D. Whitaker, a retired veterinarian of Farmville, Virginia. The puzzles are beautifully made of fine woods (Whitaker turns them out in a basement workshop), and many of them are enormously complicated and diabolically clever. A typical puzzle is a box with an opening at the top into which you drop a steel ball. The object is to get the ball out through a hole in the side of the box. One is allowed to manipulate the box in any manner, provided, of course, it is not damaged or taken apart. Much more is required than just tipping the box to roll the ball through concealed passageways. Certain impediments must be removed by tapping the box in certain ways. Other barriers have to be lifted by applying magnets or blowing through small holes. Interior magnets are so placed that they grab the ball and hold it. You are unaware of this because there are dummy balls inside that you hear rattling about. On the

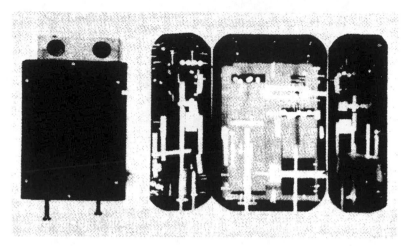

Figure 98. To solve and keep one of his puzzles (left) Grimes had to have it X-rayed (right).

outside of the box there may be wheels, levers and plungers of various types. Some of them must be manipulated a certain way to get the ball through the box; others are there just to confuse you. It may be necessary at some point to push a pin through an inconspicuous hole.

For several years Grimes and Whitaker had an arrangement whereby Grimes received a new puzzle at regular intervals. If he solved it in a month, he was permitted to keep it; otherwise, he had to buy it. In some instances, the challenge was accompanied by vigorous side bets. Once Grimes worked for almost a year on a Whitaker puzzle without cracking it. He had gone over it with a small compass to locate all concealed magnets. He had carefully probed all the openings with bent wires. The bottleneck was a plunger that had to be pushed in, but apparently some interior steel balls prevented this. Grimes correctly deduced that these balls were to be tilted out of the way, but all his attempts to do this were unsuccessful. He finally solved the puzzle by having it X-rayed (see Figure 98). The prints disclosed one large cavity into which four balls had to be rolled, and a smaller cavity into which a fifth ball had to be maneuvered. When all five balls were out of the way, the plunger yielded.

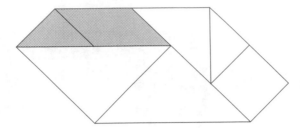

Figure 99. The elusive polygon.

The rest of the puzzle was not so difficult, though at one point it required three hands. While the right and left hands applied pressure at certain spots, another plunger, attached to a strong spring, had to be pulled out. Grimes finally managed it by tying one end of a cord to the plunger and the other end to his foot!

ANSWERS

The tangram hexagon, usually the hardest to find of the 13 possible convex tangrams, is depicted in Figure 99. The solution is unique except for the fact that the two shaded pieces may be transposed.

The peg-jumping puzzle is solved in 46 moves as follows: 10–8–7–9–12–6–3–9–15–16–10–8–9–11–14–12–6–5–8–2–1–7–9–11–17–16–10–13–12–6–4–7–9–10–8–2–3–9–15–12–6–9–11–10–8–9. At the halfway point the black and white counters form a symmetrical pattern on the board. The remaining moves repeat in reverse order the pattern of moves in the first half.

Many readers sent other elegant solutions in 46 moves. James R. Lawson of Schenectady, New York, age 14, found 48 basically different 46-move solutions.

POSTSCRIPT

As far as I know, my short article in *Hobbies* (see the bibliography) was the first article published about collecting mechanical puzzles. At the time I wrote it, at age 20, I had a small collection which I later gave to Jay Findley Christ, a professor in the University of Chicago's business school, and a noted Sherlockian scholar. Christ owned a collection larger than mine. He in turn sold it to Jules Traub, a magician who founded Fun, Inc., a firm that makes novelties.

Today, collections of mechanical puzzles, here and in Canada and abroad, far exceed the collection I wrote about in this chapter. Among living collectors Jerry Slocum, of Beverly Hills, California, is surely the champion. His magnificent collection is so vast that he built a two-story house to contain it. The house is without windows so Jerry can control temperature, humidity, and light – controls needed to preserve his treasures.

Slocum has based several beautiful volumes on his collection, starting with *Puzzles Old and New*, written with Jack Botermans (Washington University Press, 1986). This was followed by *The New Book of Puzzles, Ingenious and Diabolical Puzzles*, and two books published by Klutz Press. Jerry's *Tangram Book* (Sterling, 2003) is the first accurate history of tangrams.

The 15 Puzzle: How It Drove the World Crazy, by Slocum and Dic Sonneveld, published by the Slocum Foundation in 2006, is a history of the 14–15 puzzle that became a craze in the 1880s. Sam Loyd, in addition to being a great puzzle maker, was also something of a scoundrel. He falsely claimed credit for inventing the 14–15 puzzle, but actually had nothing to do with it aside from offering a prize for a solution. He was safe in the offer because it could be proved that the puzzle was unsolvable. Jerry's book is a thorough history of the craze, including a revelation of the true inventor.

In 2005, Jerry gave his collection of more than 3,000 puzzles to Indiana University's library, along with an equal number of books on puzzles. The collection was opened to the public in July 2006 with an exhibit attended by puzzle buffs from around the world. Margaret Wertheim wrote a report on the opening that ran in *The New York Times* science section on June 25, 2006. Another account of the exhibit, by Julie Mahomed, appeared in the *Indiana Daily Student* (August 2). Still another, by Nicole Kauffman, was in the Bloomington, Indiana *Herald-Times* (August 3, 2006). A handsome booklet by Julian Hinchcliff honoring the exhibit was published by the university's Lilly Library.

Slocum's latest book, *Het Ultieme Puzzelboek* (Terra, 2007), coauthored with Jack Botermans, is a beautifully illustrated account of the finest puzzles in his collection, their histories, and how to make some of them. Its publication in English is planned for 2009.

At the time I write, dozens of fantastic new mechanical puzzles appear every year on the market, especially in Japan. Some are solved only by first spinning the puzzle. One wooden puzzle is hard to take apart unless you roll it on a rug; then all the pieces scatter. Many new puzzles involving cords are topological. There are boxes on sale that open only after you make a dozen or more moves in a certain order! A Dutch periodical in English, *Cubism for Fun*, is devoted entirely to new mechanical puzzles.

When I was a boy I once read in a detective magazine a story, presumably one of a series, about a detective who collected mechanical puzzles. He was able to transfer his ability to solve such puzzles over to the solving of crimes. I no longer recall the author or the magazine's name. Can any reader help run this down?

I devoted two *Scientific American* columns to tangrams. They are reprinted in my Book 12.

BIBLIOGRAPHY

Puzzles Old and New. Professor Hoffmann (pseudonym of Angelo Lewis). Frederick Warne and Company, 1893. Contains pictures and descriptions of almost all the mechanical puzzles sold in England during the author's time.

Miscellaneous Puzzles. A. Duncan Stubbs. Frederick Warne and Company, 1931. Includes many unusual mechanical puzzles that can be constructed by the reader.

"A Puzzling Collection." Martin Gardner in *Hobbies* (September 1934): 8.

100 Puzzles: How to Make and How to Solve Them. Anthony S. Filipiak. A. S. Barnes and Company, 1942.

"A Theorem on the Tangram." Fu Traing Wang and Chuan-Chih Hsiung in *The American Mathematical Monthly* 49 (November 1942): 596–599.

"Making and Solving Puzzles." Jerry Slocum in *Science and Mechanics* (October 1955): 121–126.

"Classification of Mechanical Puzzles and Physical Objects Related to Puzzles." James Dalgety and Edward Hordern in *The Mathemagician and Pied Puzzler, eds.* Elwyn Berlekamp and Tom Rodgers. A K Peters, 1999.

"Diabolical Puzzles From Japan." Nob Yoshigahara, Minoru Abe, and Mineyuki Uyematsu in *Puzzler's Tribute, eds.* David Wolfe and Tom Rodgers. A K Peters, 2002.

"The Box Wizard." Tom Cutrofello in *Games* (April 2007). An account of the puzzle boxes made by Kagen Schaefer, some of which require more than 50 moves to open!

"The Precision Puzzlemaker." Tom Cutrofello in *Games* (July 2007): 66–67. On mechanical puzzles created by Michael Toulouzas.

The following four chapters all appear in *Tribute to a Mathemagician*, eds. Barry Cipra et al. A K Peters, 2005.

"Nob Yoshigahara." Jerry Slocum. A tribute to one of Japan's most creative inventors of mechanical puzzles.

"Nobuyuki Yoshigahara (1936–2004)." Bill Ritchie. Another tribute.

"Chinese Ceramic Puzzle Vases." Norman L. Sandfield.

"Mongolian Interlocking Puzzles." Jerry Slocum and Frans de Vreugd.

CHAPTER NINETEEN

Probability and Ambiguity

CHARLES SANDERS PIERCE once observed that in no other branch of mathematics is it so easy for experts to blunder as in probability theory. History bears this out. Leibniz thought it just as easy to throw 12 with a pair of dice as to throw 11. Jean le Rond d'Alembert, the great eighteenth-century French mathematician, could not see that the results of tossing a coin three times are the same as tossing three coins at once, and he believed (as many amateur gamblers persist in believing) that after a long run of heads, a tail is more likely.

Today, probability theory provides clear, unequivocal answers to simple questions of this sort, but only when the experimental procedure involved is precisely defined. A failure to do this is a common source of confusion in many recreational problems dealing with chance. A classic example is the problem of the broken stick. If a stick is broken at random into three pieces, what is the probability that the pieces can be put together in a triangle? This cannot be answered without additional information about the exact method of breaking to be used.

One method is to select, independently and at random, two points from the points that range uniformly along the stick, and then to break the stick at these two points. If this is the procedure to be followed, the answer is 1/4, and there is a neat way of demonstrating it with a geometrical diagram. We draw an equilateral triangle and then connect the midpoints of the sides to form a smaller shaded equilateral triangle in the center (see Figure 100). If we take any point in the large triangle and draw perpendiculars to the three sides, the sum of these three lines will be constant and equal to the altitude of the large triangle. When this point, like point A, is *inside* the shaded triangle, no one of the three perpendiculars will

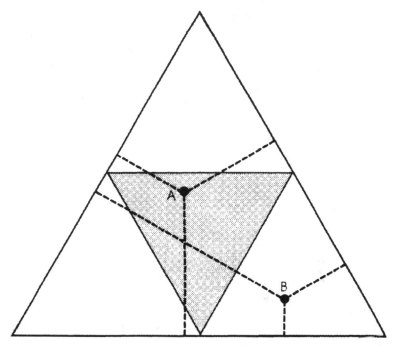

Figure 100. If a stick is broken in three pieces, the probability is 1/4 that they will form a triangle. (Artist: Amy Kasai)

be longer than the sum of the other two. Therefore the three line-segments will form a triangle. On the other hand, if the point, like point B, is *outside* the shaded triangle, one perpendicular is sure to be longer than the sum of the other two, and consequently no triangle can be formed with the three line segments.

We now have a neat geometrical analogy to the problem of the broken stick. The sum of the three perpendiculars corresponds to the length of the stick. Each point on the large triangle represents a unique way of breaking the stick, the three perpendiculars corresponding to the three broken pieces. The probability of breaking the stick favorably is the same as the probability of selecting a point at random and finding that its three perpendiculars will form a triangle. As we have seen, this happens only when the point is inside the shaded triangle. Since this area is one-fourth the total area, the probability is 1/4.

Suppose, however, that we interpret in a different way the state-ment "break a stick at random into three pieces." We break the stick at random, select randomly one of the two pieces, and break that piece at random. What are the chances that the three pieces will form a triangle?

The same diagram will provide the answer. If after the first break we choose the smaller piece, no triangle is possible. What happens when we pick the larger piece? Let the vertical perpendicular in the diagram represent the smaller piece. In order for this line to be smaller than the sum of the other two perpendiculars, the point where the lines meet cannot be inside the small triangle at the top of the diagram. It must range uniformly over the lower three triangles. The shaded triangle continues to represent favorable points, but now it is only one-third the area under consideration. The chances, therefore, are 1/3 that when we break the larger piece, the three pieces will form a triangle. Since our chance of picking the larger piece is 1/2, the answer to the original question is the product of 1/2 and 1/3, or 1/6.

Geometrical diagrams of this sort must be used with caution because they too can be fraught with ambiguity. For example, con-sider this problem discussed by Joseph Bertrand, a famous French mathematician. What is the probability that a chord drawn at ran-dom inside a circle will be longer than the side of an equilateral tri-angle inscribed in the circle?

We can answer as follows. The chord must start at some point on the circumference. We call this point A, and then draw a tan-gent to the circle at A, as shown in the top illustration of Figure 101. The other end of the chord will range uniformly over the circum-ference, generating an infinite series of equally probable chords, samples of which are shown on the illustration as broken lines. It is clear that only those chords that cut across the triangle are longer than the side of the triangle. Since the angle of the triangle at A is 60 degrees, and since all possible chords lie within a 180-degree range, the chances of drawing a chord larger than the side of the triangle must be 60/180, or 1/3.

Now let us approach the same problem a bit differently. The chord we draw must be perpendicular to one of the circle's diam-eters. We draw the diameter, then add the triangle as shown in the

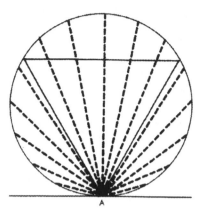

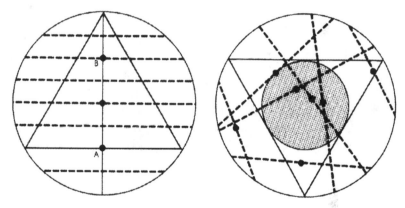

Figure 101. Probability that random chord is longer than side of inscribed equilateral triangle is proved to be 1/3 (top), 1/2 (left), and 1/4 (right). (Artist: Amy Kasai)

illustration at bottom left of Figure 101. All chords perpendicular to this diameter will pass through a point that ranges uniformly along the diameter. Samples of these chords are again shown as broken lines. It is not hard to prove that the distance from the center of the circle to A is half the radius. Let B mark the midpoint on the other side of the diameter. It is now easy to see that only those chords crossing the diameter between A and B will be longer than the side of the triangle. Since AB is half the diameter, we obtain an answer to our problem of 1/2.

Here is a third approach. The midpoint of the chord will range uniformly over the entire space within the circle. A study of the illustration at bottom right of Figure 101 will convince you that only chords whose midpoints lie within the smaller shaded circle are longer than the side of the triangle. The area of the small circle is exactly one-fourth the area of the large circle, so the answer to our problem now appears to be 1/4.

Which of the three answers is right? Each is correct in reference to a certain mechanical procedure for drawing a random chord. Examples of the three procedures are as follows:

1. Two spinners are mounted at the center of a circle. They rotate independently. We spin them, mark the two points at which they stop, and connect the points with a straight line. The probability that this line will be longer than the side of the inscribed triangle is 1/3.

2. A large circle is chalked on the sidewalk. We roll a broom handle toward it, from a distance of 50 feet, until the handle stops somewhere on the circle. The probability that it will mark a chord longer than the side of the triangle is 1/2.

3. We paint a circle with molasses and wait until a fly lights on it; then we draw the chord on which the fly is the midpoint. The probability that this chord is longer than the side of the triangle is 1/4.

Each of these procedures is a legitimate method of obtaining a "random chord." The problem as originally stated, therefore, is ambiguous. It has no answer until the meaning of "draw a chord at random" is made precise by a description of the procedure to be followed. Apparently nothing resembling any of the three procedures is actually adopted by most people when they are asked to draw a random chord. In an interesting unpublished paper entitled "The Human Organism as a Random Mechanism" Oliver L. Lacey, professor of psychology at the University of Alabama, reports on a test that showed the probability to be much better than 1/2 that a subject would draw a chord longer than the side of the inscribed triangle.

Another example of ambiguity arising from a failure to specify the randomizing procedure appears in Chapter 14, Problem 2. Readers were told that Mr. Smith had two children, at least one of whom was a boy, and were asked to calculate the probability that both were boys. Many readers correctly pointed out that the answer depends on the procedure by which the information "at least one is a boy" is obtained. If from all families with two children, at least one of whom is a boy, a family is chosen at random, then the answer is 1/3. But there is another procedure that leads to exactly the same statement of the problem. From families with two children, one family is selected at random. If both children are boys, the informant says "at least one is a boy." If both are girls, he says "at least one is a girl." And if both sexes are represented, he picks a child at random and says "at least one is a . . . ," naming the child picked. When *this* procedure is followed, the probability that both children are of the same sex is clearly 1/2. (This is easy to see because the informant makes a statement in each of the four cases – BB, BG, GB, GG – and in half of these cases both children are of the same sex.)

The following wonderfully confusing little problem involving three prisoners and a warden is even more difficult to state unambiguously. Three men – A, B, and C – were in separate cells under sentence of death when the governor decided to pardon one of them. He wrote their names on three slips of paper, shook the slips in a hat, drew out one of them, and telephoned the warden, requesting that the name of the lucky man be kept secret for several days. Rumor of this reached prisoner A. When the warden made his morning rounds, A tried to persuade the warden to tell him who had been pardoned. The warden refused.

"Then tell me," said A, "the name of one of the others who will be executed. If B is to be pardoned, give me C's name. If C is to be pardoned, give me B's name. And if I'm to be pardoned, flip a coin to decide whether to name B or C."

"But if you see me flip the coin," replied the wary warden, "you'll know that you're the one pardoned. And if you see that I don't flip a coin, you'll know it's either you or the person I don't name."

"Then don't tell me now," said A. "Tell me tomorrow morning."

The warden, who knew nothing about probability theory, thought it over that night and decided that if he followed the procedure suggested by A, it would give A no help whatever in estimating his survival chances. So next morning he told A that B was going to be executed.

After the warden left, A smiled to himself at the warden's stupidity. There were now only two equally probable elements in what mathematicians like to call the "sample space" of the problem. Either C would be pardoned or himself, so by all the laws of conditional probability, his chances of survival had gone up from 1/3 to 1/2.

The warden did not know that A could communicate with C, in an adjacent cell, by tapping in code on a water pipe. This A proceeded to do, explaining to C exactly what he had said to the warden and what the warden had said to him. C was equally overjoyed with the news because he figured, by the same reasoning used by A, that his own survival chances had also risen to 1/2.

Did the two men reason correctly? If not, how should each have calculated his chances of being pardoned?

ADDENDUM

In giving the second version of the broken stick problem, I could hardly have picked a better illustration of the ease with which experts can blunder on probability computations, and the dangers of relying on a geometrical diagram. My solution was taken from William A. Whitworth's *DCC Exercises in Choice and Chance*, Problem 677; the same answer will be found in many other older textbooks on probability. It is entirely wrong!

In the first version of the problem, in which the two breaking points are simultaneously chosen, the representative point on the diagram ranges uniformly over the large triangle, permitting a comparison of areas to obtain a correct answer. In the second version, in which the stick is broken, then the larger piece is broken, Whitworth assumed that the point on the diagram ranged uniformly over the three lower triangles. It doesn't. There are more points within the central triangle than in the other two.

Let the length of the stick be 1 and x be the length of the smallest piece after the first break. To obtain pieces that will form a triangle, the larger segment must be broken within a length equal to $1 - x$. Therefore the probability of obtaining a triangle is $x/(1 - x)$. We now have to average all values of x, from 0 to 1/2, to obtain a value for this expression. It proves to be $-1 + 2 \log 2$, or 0.386. Since the probability is 1/2 that the larger piece will be picked for breaking, we multiply 0.386 by 1/2 to obtain 0.193, the answer to the problem. This is a trifle larger than 1/6, the answer obtained by following Whitworth's reasoning.

A large number of readers sent very clear analyses of the problem. In the preceding summary, I followed a solution sent by Mitchell P. Marcus of Binghamton, NY. Similar solutions were received from Edward Adams, Howard Grossman, Robert C. James, Gerald R. Lynch, G. Bach and R. Sharp, David Knaff, Norman Geschwind, and Raymond M. Redheffer. Professor Redheffer, at the University of California, is coauthor (with Ivan S. Sokolnikoff) of *Mathematics of Physics and Modern Engineering* (McGraw-Hill, 1958), in which will be found (page 636) a full discussion of the problem. See also *Ingenious Mathematical Problems and Methods* by L. A. Graham (Dover, 1959, Problem 32) for other methods of solving the problem's first version.

Frederick R. Kling, John Ross, and Norman Cliff, all with the Educational Testing Service, Princeton, NJ, also sent a correct solution of the problem's second version. At the close of their letter they asked which of the following three hypotheses was most probable:

1. Mr. Gardner honestly blundered.
2. Mr. Gardner deliberately blundered in order to test his readers.
3. Mr. Gardner is guilty of what is known in the mathematical world as keeping up with the d'Alemberts.

The answer: number three.

ANSWERS

The answer to the problem of the three prisoners is that A's chances of being pardoned are 1/3, and that C's chances are 2/3.

Regardless of who is pardoned, the warden can give A the name of a man, other than A, who will die. The warden's statement therefore has no influence on A's survival chances; they continue to be 1/3.

The situation is analogous to the following card game. Two black cards (representing death) and a red card (the pardon) are shuffled and dealt to three men: A, B, C (the prisoners). If a fourth person (the warden) peeks at all three cards, then turns over a black card belonging to either B or C, what is the probability that A's card is red? There is a temptation to suppose it is 1/2 because only two cards remain face-down, one of which is red. But since a black card can always be shown for B or C, turning it over provides no information of value in betting on the color of A's card.

This is easy to understand if we exaggerate the situation by letting death be represented by the Ace of Spades in a full deck. The deck is spread, and A draws a card. His chance of avoiding death is 51/52. Suppose now that someone peeks at the cards, and then turns face up 50 cards that do not include the Ace of Spades. Only two face-down cards are left, one of which must be the Ace of Spades, but this obviously does not lower A's chances to 1/2. It doesn't because it is always possible, if one looks at the faces of the 51 cards, to find 50 that do not include the Ace of Spades. Finding them and turning them face up, therefore, has no effect on A's chances. Of course if 50 cards are turned over at random, and none prove to be the Ace of Spades, then the chance that A drew the death card *does* rise to 1/2.

What about prisoner C? Since either A or C must die, their respective probabilities for survival must add up to 1. A's chances to live are 1/3; therefore C's chances must be 2/3. This can be confirmed by considering the four possible elements in our sample space, and their respective initial probabilities:

1. C is pardoned, warden names B (probability 1/3).
2. B is pardoned, warden names C (probability 1/3).
3. A is pardoned, warden names B (probability 1/6).
4. A is pardoned, warden names C (probability 1/6).

Only cases 1 and 3 apply when it becomes known that B will die. The chances that it is case 1 are 1/3, or twice the chances (1/6) that it is case 3, so C's survival chances are two to one, or 2/3. In the card-game model, this means that there is a probability of 2/3 that C's card is red.

This problem of the three prisoners brought a flood of mail, pro and con: Happily, all objections proved groundless. Sheila Bishop of East Haven, CT, sent the following well-thought-out analysis:

Sirs:
  I was first led to the conclusion that A's reasoning was incorrect by the following paradoxical situation. Suppose the original conversation between A and the warden had taken place in the same way, but now suppose that just as the warden was approaching A's cell to tell him that B would be executed, the warden fell down a manhole or was in some other way prevented from delivering the message.
  A could then reason as follows: "Suppose he was about to tell me that B would be executed. Then my chance of survival would be 1/2. If, on the other hand, he was going to tell me that C would be executed, then my chances would still be 1/2. Now I know as a certain fact that he would have told me one of those two things; therefore, either way, my survival chances are bound to be 1/2." Following this line of thought shows that A could have figured his chances to be 1/2 without ever asking the warden anything!
  After a couple of hours I finally arrived at this conclusion: Consider a large number of trios of prisoners all in this same situation, and in each group let A be the one who talks to the warden. If there are $3n$ trios altogether, then in n of them A will be pardoned, in n B will be pardoned, and in n C will be pardoned. There will be $3n/2$ cases in which the warden will say, "B will be executed." In n of these cases C will go free and in $n/2$ cases A will go free; C's chances are twice as good as A's. Hence A's and C's chances of survival are 1/3 and 2/3 respectively. . . .

Lester R. Ford, Jr., and David N. Walker, both with the Arizona office of General Analysis Corporation, felt that the warden has been unjustly maligned:

Sirs:
  We are writing to you on behalf of the warden, who is a political appointee and therefore unwilling to enter into controversial matters in his own behalf.
  You characterize him in a slurring manner as "The warden, who knew nothing about probability theory, ..." and I feel that a grave injustice is being done. Not only are you incorrect (and possibly

libelous), but I can personally assure you that his hobby for many years has been mathematics, and in particular, probability theory. His decision to answer A's question, while based on a humanitarian attempt to brighten the last hours of a condemned man (for, as we all now know, it was C who received the pardon), was a decision completely compatible with his instructions from the governor.

The only point on which he is open to criticism (and on this he has already been reprimanded by the governor) is that he was unable to prevent A from communicating with C, thereby permitting C to more accurately estimate *his* chances of survival. Here too, no great damage was done, since C failed to make proper use of the information.

If you do not publish both a retraction and an apology, we shall feel impelled to terminate our subscription.

ADDENDUM

The problem of the two boys, as I said, must be very carefully stated to avoid ambiguity that prevents a precise answer. In my *Aha, Gotcha* I avoided ambiguity by imagining a lady who owned two parrots – one white, one black. A visitor asks the owner, "Is one bird a male?" The owner answers yes. The probability both parrots are male is 1/3. Had the visitor asked, "Is the dark bird a male?" a yes answer would have raised the probability that both birds are male to 1/2.

Richard E. Bedient, a mathematician at Hamilton College, described the prisoner's paradox in a poem that appeared in *The American Mathematical Monthly*, Vol. 101, March 1994, page 249:

The Prisoner's Paradox Revisited

Awaiting the dawn sat three prisoners wary,
A trio of brigands named Tom, Dick and Mary.
Sunrise would signal the death knoll of two,
Just one would survive, the question was who.

Young Mary sat thinking and finally spoke.
To the jailer she said, "You may think this a joke
But it seems that my odds of surviving 'til tea,
Are clearly enough just one out of three.

But one of my cohorts must certainly go,
Without question, that's something I already know.
Telling the name of one who is lost,

Can't possibly help me. What could it cost?"

The shriveled old jailer himself was no dummy,
He thought, "But why not?" and pointed to Tommy.
"Now it's just Dick and I" Mary chortled with glee,
"One in two are my chances, and not one in three!"

Imagine the jailer's chagrin, that old elf,
She'd tricked him, or had she? Decide for yourself.

When I introduced the three prisoners paradox in my October 1959 column, I received a raft of letters from mathematicians who believed my solution was invalid. The number of such letters, however, was small compared to the thousands of letters Marilyn vos Savant received when she gave a version of the problem in her popular *Parade* column for September 9, 1990.

Ms. Savant's version of the paradox was based on a then-popular television show called *Let's Make a Deal*, hosted by Monty Hall. Imagine three doors, Marilyn wrote, to three rooms. Behind one door is a prize car. Behind each of the other two doors is a goat. A guest on the show is given a chance to win the prize by selecting the door with the car. If she chooses at random, clearly the probability she will select the prize door is 1/3. Now suppose, that after the guest's selection is voiced, Monty Hall, who knows what is behind each door, opens one door to disclose a goat. Two closed doors remain. One might reason that because the car is now behind one of just two doors, the probability the guest had chosen the correct door has risen to 1/2. Not so! As Marilyn correctly stated, it remains 1/3. Because Monty can always open a door with a goat, his opening such a door conveys no new information that alters the 1/3 probability.

Now comes an even more counterintuitive result. If the guest switches her choice from her initial selection to the other closed door, her chances of winning rise to 2/3. This should be obvious if one grants that the probability remains 1/3 for the first selection. The car must be behind one of the two doors, therefore the probabilities for each door must add to 1, or certainty. If one door has a probability of 1/3 being correct, the other door must have a 2/3 probability.

Marilyn was flooded with letters from irate readers, many accusing her of being ignorant of elementary probability theory and many from professional mathematicians. So awesome was the mail, and so controversial, that *The New York Times*, on July 21, 1991, ran a front page, lengthy feature about the flap. The story, written by John Tierney, was titled "Behind Monty Hall's Doors: Puzzle, Debate and Answer?" (See also letters about the feature in *The New York Times*, August 11, 1991.)

The red-faced mathematicians, who were later forced to confess they were wrong, were in good company. Paul Erdös, one of the world's greatest mathematicians, was among those unable to believe that switching doors doubled the probability of success. Two recent biographies of the late Erdös reveal that he could not accept Marilyn's analysis until his friend Ron Graham patiently explained it to him.

The Monty Hall Problem, as it came to be known, generated many articles in mathematical journals. I list some of them in this chapter's bibliography.

It is hard to believe, but apparently my 1959 mention of the three-prisoner problem, which is the same as the notorious Monty Hall Problem, was the first appearance of this problem in print. I cannot now recall who first told me about the three prisoners. I soon realized how it could be modeled with three playing cards or three shells and a pea. Jason Rosenhause, a mathematician at James Madison University, has written an entire book about the problem. Titled *The Monty Hall Problem*, it is scheduled for publication in 2009 by Oxford University Press. The book is a marvelous history and in-depth analysis of the problem, and its many variations and generalizations. Its bibliography lists 100 references!

BIBLIOGRAPHY

THE THREE PRISONERS PARADOX

"The Problem of the Three Prisoners." D. H. Brown in *The Mathematics Teacher* (February 1966): 181–182.

"A Paradox in Probability Theory." N. Starr in *The Mathematics Teacher* (February 1973): 166–168.

"Intuitive Reasoning About Probability: Theoretical and Experiential Analysis of Three Prisoners." S. Shimojo and S. Ichikano in *Cognition* 32 (1989): 1–24.

"Erroneous Beliefs in Estimating Posterior Probability." S. Ichikawa and H. Takeichi in *Bahaviormetrika* 27 (1990): 59–73.
"A Closer Look at the Probabilities of the Notorious Three Prisoners." R. Falk in *Cognition* 43 (1992): 197–223.

THE MONTY HALL PROBLEM

"Ask Marilyn." M. vos Savant in *Parade*, September 9, 1990; December 2, 1990; February 17, 1991; July 7, 1991; September 8, 1991; October 13, 1991; January 5, 1992; January 26, 1992.
"Ask Marilyn: The Mathematical Controversy in *Parade* Magazine." A. Lo Bello in *Mathematical Gazette* 75 (October 1991): 271–272.
"The Car and the Goats." L. Gillman in *American Mathematical Monthly* 99 (1992): 3–7.
"Fallacies, Flaws, and Flimflams." E. Barbeau in *The College Mathematics Journal* 26 (May 1995): 132–184.
"A Tale of Two Goats . . . and a Car." A. H. Bohl, M. J. Liberatore, and R. L. Nydick in *Journal of Recreational Mathematics* 27 (1995): 1–9.
"Generalizing Monty's Dilemma." J. P. Georges and T. V. Craine in *Quantum* (March/April 1995): 17–21.
"A Mathematical Excursion: From the Three-door Problem to a Cantortype Set." J. Paradis, P. Viader, and L. Bibiloni in *American Mathematical Monthly* 106 (March 1999): 241–251.
"Monte's Dilemma: Should You Stick or Switch?" J. M. Shaughnessy and T. Dick in *The Mathematics Teacher* (April 1991): 252–256.

CHAPTER TWENTY

The Mysterious Dr. Matrix

NUMEROLOGY, THE STUDY of the mystical significance of numbers, has a long, complicated history that includes the ancient Hebrew cabalists, the Greek Pythagoreans, Philo of Alexandria, the Gnostics, many distinguished theologians, and those Hollywood numerologists who prospered in the 1920s and 1930s by devising names (with proper "vibrations") for would-be movie stars. I must confess that I have always found this history rather boring. Thus when a friend of mine suggested that I get in touch with a New York numerologist who calls himself Dr. Matrix, I could hardly have been less interested.

"But you'll find him very amusing," my friend insisted. "He claims to be a reincarnation of Pythagoras, and he really does seem to know something about mathematics. For example, he pointed out to me that 1960 had to be an unusual year because 1,960 can be expressed as the sum of two squares – 14^2 and 42^2 – and both 14 and 42 are multiples of the mystic number 7."

I made a quick check with pencil and paper. "By Plato, he's right!," I exclaimed. "He might be worth talking to at that."

I telephoned for an appointment, and several days later a pretty secretary with dark, almond-shaped eyes ushered me into the doctor's inner sanctum. Ten huge numerals from 1 to 10, gleaming like gold, were hanging on the far wall behind a long desk. They were arranged in the triangular pattern made commonplace today by the arrangement of bowling pins, but which the ancient Pythagoreans viewed with awe as the "Holy Tetractys." A large dodecahedron on the desk bore a calendar for each month of the new year on each of its 12 sides. Soft organ music was coming from a hidden loudspeaker.

Figure 102. Dr. Matrix's alphabet circle.

Dr. Matrix entered the room through a curtained side door; he was a tall, bony figure with a prominent nose and bright, penetrating eyes. He motioned me into a chair. "I understand you write for *Scientific American*," he said with a crooked smile, "and that you're here to inquire about my methods rather than for a personal analysis."

"That's right," I said.

The doctor pushed a button on a side wall, and a panel in the woodwork slid back to reveal a small blackboard. On the blackboard were chalked the letters of the alphabet, in the form of a circle that joined Z to A (see Figure 102). "Let me begin," he said, "by explaining why 1960 is likely to be a favorable year for your magazine." With the end of a pencil he began tapping the letters, starting with A and proceeding around the circle until he counted 19. The 19th letter was S. He continued around the circle, starting with the count of 1 on T, and counted up to 60. The count ended on A. S and A, he pointed out, are the initials of *Scientific American*.

"I'm not impressed," I said. "When there are thousands of different ways that coincidences like this can arise, it becomes extremely probable that with a little effort you can find at least one."

"I understand," said Dr. Matrix, "but don't be too sure that's the whole story. Coincidences like this occur far more often than can be justified by probability theory. Numbers, you know, have a mysterious life of their own." He waved his hand toward the gold numerals on the wall. "Of course those are not numbers. They're only symbols for numbers. Wasn't it the German mathematician Leopold Kronecker who said: 'God created the integers; all the rest is the work of man'?"

"I'm not sure I agree with that," I said, "but let's not waste time on metaphysics."

"Quite right," he replied, seating himself behind the desk. "Let me cite a few examples of numerological analysis that may interest your readers. You've heard, perhaps, the theory that Shakespeare worked secretly on part of the King James translation of the Bible?"

I shook my head.

"To a numerologist, there's no doubt about it. If you turn to the 46th Psalm you'll find that its 46th word is 'shake.' Count back to the 46th word from the end of the same psalm [the world *selah* at the end is not part of the psalm] and you read the word 'spear.'"

"Why 46?" I asked, smiling.

"Because," said Dr. Matrix, "when the King James Authorized Version was completed in 1610, Shakespeare was exactly 46 years old."

"Not bad," I said as I scribbled a few notes. "Any more?"

"Thousands," said Dr. Matrix. "Consider the case of Richard Wagner and the number 13. There are 13 letters in his name. He was born in 1813. Add the digits of this year and the sum is 13. He composed 13 great works of music. *Tannhäuser*, his greatest work, was completed on April 13, 1845, and first performed on March 13, 1861. He finished *Parsifal* on January 13, 1882. *Die Walküre* was first performed in 1870 on June 26, and 26 is twice 13. *Lohengrin* was composed in 1848, but Wagner did not hear it played until 1861, exactly 13 years later. He died on February 13, 1883. Note that the first and last digits of this year also form 13. These are only a few of the many important 13's in Wagner's life."

Dr. Matrix waited until I had finished writing; then he continued. "Important dates are never accidental. The atomic age began in 1942, when Enrico Fermi and his colleagues achieved the first nuclear chain reaction. You may have read in Laura Fermi's biography of her husband how Arthur Compton telephoned James Conant to report the news. Compton's first remark was: 'The Italian navigator has reached the New World.' Did it ever occur to you that if you switch the middle digits of 1942, it becomes 1492, the year that Columbus, an earlier Italian navigator, discovered the New World?"

"Never," I answered.

"The life of Kaiser Wilhelm I is numerologically interesting," he went on. "In 1849 he crushed the socialist revolution in Germany. The sum of the digits in this date is 22. Add 22 to 1849 and you get 1871, the year Wilhelm was crowned emperor. Repeat this procedure with 1871, and you arrive at 1888, the year of his death. Repeat once more and you get 1913, the last year of peace before World War I destroyed his empire. Unusual date patterns are common in the lives of all famous men. Is it coincidence that Raphael, the great painter of sacred scenes, was born on April 6 and died on April 6, and that both dates fell on Good Friday? Why is evolution a key to the philosophies of both John Dewey and Henri Bergson? Because both men were born in 1859, the year Darwin's *Origin of Species* was published. Do you think it accidental that Houdini, the lover of mystery, died on October 31, the date of Halloween?"

"Could be," I murmured.

The doctor shook his head vigorously. "I suppose you'll think it coincidental that in the library's Dewey decimal system the classification for books on number theory is 512.81."

"Is there something unusual about that?"

"The number 512 is 2 to the ninth power and 81 is 9 to the second power. But here's something even more remarkable. First, 11 plus 2 minus 1 is 12. Let me show you how this works out with letters." He moved to the blackboard and chalked on it the word ELEVEN. He added TWO to make ELEVEN-TWO, then he erased the letters of ONE, leaving ELEVTW. "Rearrange those six letters," he said, "and they spell TWELVE."

I dabbed at my forehead with my handkerchief. "Do you have any opinion about 666," I asked, "the so-called Number of the Beast

[Revelation 13:18]? I recently came across a book called *Our Times and Their Meaning*, by a Seventh-Day Adventist named Carlyle B. Haynes. He identified the number with the Roman Catholic Church by adding up all the Roman numerals in one of the Latin titles of the Pope: VICARIUS FILII DEI. It comes to exactly 666." [V = 5, I = 1, C = 100, I = 1, U = 5, I = 1, L = 50, I = 1, I = 1, D = 500, I = 1. U is taken as V because that is how it used to be written.]

"I could talk for hours about 666," the doctor said with a heavy sigh. "This particular application of the Beast's number is quite old. Of course it's easy for a skillful numerologist to find 666 in any name. In fact, if you add the Latin numerals in the name ELLEN GOULD WHITE, the inspired prophetess who founded Seventh-Day Adventism – counting W as a 'double U' or two V's – it also adds up to 666. [L = 50, L = 50, U = 5, L = 50, D = 500, W = 10, I = 1.] Tolstoy's *War and Peace* [Volume III, Part 1, Chapter 19] has a neat method of extracting 666 from L'EMPEREUR NAPOLEON. When the prime minister of England was William Gladstone, a political enemy wrote GLADSTONE in Greek, added up the Greek numerals in the name and got 666. HITLER adds up neatly to the number if we use a familiar code in which A is 100, B is 101, C is 102, and so on."

"I think it was the mathematician Eric Temple Bell," I said, "who discovered that 666 is the sum of the integers from 1 to 36, the numbers on a roulette wheel."

"True," said Dr. Matrix. "And if you put down from right to left the first six Roman numerals, in serial order, you get this." He wrote DCLXVI (which is 666) on the blackboard.

"But what does it all mean?" I asked.

Dr. Matrix was silent for a moment. "The true meaning is known only to a few initiates," he said unsmilingly. "I'm afraid I can't reveal it at this time."

"Would you be willing to comment on the coming presidential campaign?" I asked. "For instance, will Nixon or Rockefeller get the Republican nomination?"

"That's another question I prefer not to answer," he said, "but I would like to call your attention to some curious counterpoint involving the two men. 'Nelson' begins and ends with N. 'Rockefeller' begins and ends with R. Nixon's name has the same pattern in reverse. 'Richard' begins and almost ends with R. 'Nixon' begins and ends with N. Do you know when and where Nixon was born?"

"No," I said.

"At Yorba Linda, California – in January, 1913." Dr. Matrix turned back to the blackboard and wrote this date as 1–1913. He added the digits to get 15. On the circular alphabet he circled Y, L, and C, the initials of Nixon's birthplace, then he counted from each letter to the 15th letter from it clockwise to obtain NAR, the initials of Nelson Aldrich Rockefeller! "Of course," he added, "of the two men, Rockefeller has the better chance to be elected."

"How is that?"

"His name has a double letter. You see, because of the number 2 in 20th century, every president of this century must have a double letter in his name, like the OO in Roosevelt and the RR in Harry Truman."

"Ike doesn't have a double letter," I said.

"Eisenhower is the one exception so far. We must remember, however, that he ran twice against Adlai Ewing Stevenson, who also lacks the double letter. Ike's double initials 'D. D.' were sufficient to give him the advantage."

I glanced toward the blackboard. "Any other uses for that circular alphabet?"

"It has many uses," he replied. "Let me give you a recent example. The other day a young man from Brooklyn came to see me. He had renounced a vow of allegiance to a gang of hoodlums and he thought he ought to leave town to avoid punishment by gang members. Could I tell him by numerology, he wanted to know, where he should go? I convinced him he should go nowhere by taking the word ABJURER [one who renounces] and substituting for each letter the letter directly opposite it on the alphabet circle."

Dr. Matrix drew chalk lines on the blackboard from A to N, B to O, and so on. The new word was NOWHERE. "If you think that's a coincidence," he said, "just try it with even shorter words. The odds against starting with a seven-letter word and finding a second one by this technique are astronomical."

I glanced nervously at my wrist watch. "Before I leave, could you give me a numerological problem or two that I could ask my readers to solve?"

"I'll be delighted," he said. "Here's an easy one." On my notepaper he wrote the letters: OTTFFSSENT.

"On what basis are those letters ordered?" he asked. "It's a problem I give my beginning students of Neo-Pythagoreanism. Please note that the number of letters is the same as the number of letters in the name Pythagoras."

Beneath these letters he wrote:

$$
\begin{array}{r}
FORTY \\
+\ TEN \\
+\ TEN \\
\hline
SIXTY
\end{array}
$$

"Each letter in that addition problem stands for a different digit," he explained. "There's only one solution, but it takes a bit of brain work to find it."

I pocketed my pencil and paper and stood up. Organ music continued to pour into the room. "Isn't that a Bach recording?" I asked.

"It is indeed," answered the doctor as he walked me to the door. "Bach was a deep student of our science. Have you read Leonard Bernstein's *The Joy of Music*? It has an interesting paragraph about Bach's numerological investigations. He knew that the sum of the values of BACH – taking A as 1, B as 2, and so on – is 14, a multiple of the divine 7. He also knew that the sum of his entire name, using an old German alphabet, is 41, the reverse of 14, as well as the 14th prime number when you include 1 as a prime. The piece you're hearing is *Vor deinen Thron tret' ich allhier*, a hymn in which the musical form exploits this 14–41 motif. The first phrase has 14 notes, the entire melody has 41. Magnificent harmony, don't you think? If only our modern composers would learn a little numerology, they might come as close as this to the music of the spheres!"

I left the office in a slightly dazed condition; but not too dazed to notice again on my way out that the doctor's secretary had 1 upturned nose, 2 luminous eyes, and a most interesting overall figure.

ADDENDUM

The 1960 presidential election provided a dramatic confirmation of Dr. Matrix's remarks about the law of double letters. Among the top contenders for the Democratic nomination only John Fitzgerald Kennedy had the double letter, and he won both the nomination and election.

Dr. Matrix pointed out that Enrico Fermi obtained the first chain reaction in 1942, and that reversing the 94 gives 1492, the year another Italian made a great discovery. Luis W. Alvarez, a physicist at the University of California's Radiation Laboratory, in Berkeley, carried this analysis to new numerological heights. His letter appeared in *Scientific American*, April 1960:

Sirs:

I enjoyed reading Martin Gardner's account of his visit with Dr. Matrix. When the doctor was discussing the first nuclear chain reaction, he was certainly on the right track, but because he did not work actively on the Manhattan District project, he missed some important verifications of his conclusions. He would have known, of course, that the only reason the pile was built during the war was to produce plutonium, the 94th element in the periodic system. What Dr. Matrix missed by not having Manhattan District clearance was the fact that the code designation for plutonium, all during the war, was "49." If the good doctor had had this fact available to him, he would also have pointed out that element 94 was discovered in California, the land of the 49'ers.

Since the real test of a new theory is its ability to predict new relationships which the author of the theory could not have foreseen, you have convinced me that numerology is here to stay.

ANSWERS

The letters OTTFFSSENT are the initials of the names of the cardinal numbers from one to ten.

Dr. Matrix's addition problem was originated by Alan Wayne, a high school teacher of mathematics in New York, N. Y., and first appeared in the *American Mathematical Monthly*, August–September 1947, page 413. In introducing the problem, the magazine's problem editor pointed out that a "cryptarithm," to be considered "charming," should exhibit four features:

1.   The letters should make sense.
2.   All digits should be used.

3. The solution must be unique.
4. It should be solvable by logic rather than by tedious trial and error.

Wayne's cryptarithm has all four features. The unique solution is

$$
\begin{array}{r}
29786 \\
850 \\
850 \\
\hline
31486
\end{array}
$$

Note that the sum differs in only one digit from the four-decimal value of pi.

For readers who may wonder how to go about solving a cryptarithm, I quote a letter of Monte Dernham, of San Francisco, who sent the best explanation of how Wayne's problem could be analyzed:

> The repetition of TY in the first and fourth lines necessitates zero for N and 5 for E, with unity carried to the hundreds column. The double space preceding each TEN requires that O in FORTY equal 9, with 2 carried from the hundreds column, whence I denotes the unit digit 1 in 11, with F plus 1 equal to S. This leaves 2, 3, 4, 6, 7, and 8 unassigned.
>
> Since the hundreds column (*viz.*, R plus 2T plus 1) must be equal to or greater than 22, T and R must each be greater than 5, relegating F and S to 2, 3, and 4. Now X is not equal to 3; else F and S could not be consecutive integers. Then X equals 2 or 4, which, it is readily found, is impossible if T is equal to or less than 7. Hence T equals 8, with R equal to 7 and X equal to 4. Then F equals 2 and S equals 3, leaving the remaining letter, Y, equal to 6.

POSTSCRIPT

My many later visits with Dr. Matrix and his daughter Iva are collected in Book 9 of this series. For more on 666, see my article "666 and All That" in *The New Age: Notes of a Fringe Watcher* (Prometheus, 1988), and Clifford Pickover's *A Passion for Mathematics* (Wiley, 2006), pages 73, 75, 76, 84, 90–92.

BIBLIOGRAPHY

Numerology. E. T. Bell. The Williams & Wilkins Company, 1933.

"Numerology: Old and New." Joseph Jastrow in *Wish and Wisdom*. D. Appleton-Century Company, 1935.

Medieval Number Symbolism, Its Sources, Meaning and Influence on Thought and Expression. Vincent Foster Hopper. Columbia University Press, 1938.

"The Number of the Beast." Augustus De Morgan in *A Budget of Paradoxes*, Vol. 2, pages 218–240. Dover Publications, Inc., 1954.

How to Apply Numerology. James Leigh. Bazaar, Exchange and Mart, Ltd., London, 1959. A pro-numerology book by the first editor of a British occult magazine, *Prediction*.

The Magic Numbers of the Professor. Owen O'Shea and Underwood Dudley. Mathematical Association of America, 2007.

Index

Printed in the United States
By Bookmasters